The Channel Tunnel

THE CHANNEL TUNNEL

PUBLIC POLICY, REGIONAL DEVELOPMENT AND EUROPEAN INTEGRATION

by

IAN HOLLIDAY
GERARD MARCOU
ROGER VICKERMAN

With

MICHELE BREUILLARD
MICHEL LANGRAND
CLIVE CHURCH

BELHAVEN PRESS
LONDON AND NEW YORK

First published in Great Britain in 1991 by
Belhaven Press (a division of Pinter Publishers),
25 Floral Street, London WC2E 9DS

British Library Cataloguing in Publication Data
A CIP catalogue record for this book is available from the British Library

ISBN 1 85293 150 7

For enquiries in North America please contact
PO Box 197, Irvington, NY 10553

Library of Congress Cataloging in Publication Data
Holliday, Ian.
The Channel Tunnel: public policy, regional development, and European integration/by Ian Holliday, Gérard Marcou, Roger Vickerman with Michèle Breuillard, Michel Langrand, Clive Church.
p. cm.
Includes bibliographical references and index.
ISBN 1-85293-150-7
1. Tunnels - English Channel. 2. Tunnels - English Channel - Political aspects - Great Britain. 3. Tunnels - English Channel - Political aspects - France. 4. Tunnels - English Channel - Economic aspects - Europe. 5. Europe - Economic integration. I. Marcou, Gérard. II. Vickerman, R.W. (Roger William).
HE380.G73E5445 1991 91-7434
385'.312-dc20 CIP

Typeset by Mayhew Typesetting, Bristol, England
Printed and Bound by Biddles Ltd.

Contents

Figures

Tables

Preface

The end of 1990 marked a critical point in the long history of the Channel Tunnel, with the breakthrough of the service tunnel. Though this represented a major engineering feat of a type which clearly captures the imagination, it has not been engineering which has stood in the way of the Channel Tunnel in the past. Perhaps more significant, therefore, was the fact that at the same time Eurotunnel was able to secure a satisfactory refinancing of the project to cover increases in costs. The Tunnel is, in financial and management terms, a purely private-sector project, but that it should be so was a political decision and that it should have gone ahead at this time was equally a political decision. It is ironic, therefore, that the politician most implicated in both of these decisions, Margaret Thatcher, should have ended her Prime Ministerial career only a week before the breakthrough.

Ever since the announcement that Eurotunnel had been awarded the concession to build and operate a fixed link across the Channel, an Anglo-French team has been engaged on a major academic research project. The team, based in the two universities at either end of the Tunnel, the University of Kent at Canterbury and the Université de Lille II, has concerned itself with many of the wider questions surrounding the Tunnel; this book represents a synthesis of the main strands of the research. Further results are presented in G. Marcou, Y. Luchaire and R.W. Vickerman (eds), *Le Tunnel sous la Manche entre Etats et marchés* (Presses Universitaires de Lille, 1991).

The main question which has been avoided in this book is whether the initial decision to build, taken in January 1986, was the right one. This has been taken as given. More interesting questions are why the decision was taken at that particular time, why it took the form of a private concession, how public authorities responded to the decision, what the economic impact of the Tunnel will be on adjacent regions and on patterns of regional development in Europe, and how the Tunnel fits into wider debates on transport policy and European integration generally.

We have organised the material into three basic parts. In Part One, we discuss the nature of the concession arrangement itself, and the underlying economics of the Tunnel project against the background of two centuries of debate. The purpose here is to demonstrate how the basic economics of the project interact with the political context in France and the UK, and how this interaction affected the choice of the means of implementation.

In the second, and core, part of the book, we examine policy responses to the decision to build. This section sets the Tunnel in the context of the economics of the two regions it affects most, Kent and Nord-Pas de Calais, and examines institutional and political responses in the two regions to both construction and the consequence of operation. In this section we examine hypotheses about the way in which different political attitudes and structures at local and national levels affect the way in which a nationally taken decision can be implemented, modified and ameliorated at the local level. Furthermore, the extent to which there is competition or collaboration between authorities in the two countries is examined. This section is based on a considerable programme of interviews of key actors in local government and public authorities, as well as on careful monitoring of events during the crucial phase of implementation.

In Part Three, we expand the coverage of impact to examine the way the Tunnel fits into wider debates. Transport policy is an area of increasing concern, and one where the UK often appears to be increasingly out of line with both other EC countries and the EC itself. The Tunnel provides an immediate challenge to this lack of harmonisation by linking together the British and French rail systems. Whether transport infrastructure affects economic development has been the cause of much debate. What is clear is that infrastructure is not only relevant to the immediate regions of Kent and Nord-Pas de Calais, and that there is, therefore, a question about the distribution of economic activity across a wider area of Europe to be addressed. Finally, the Tunnel has a more general European significance in terms of the process of integration. This may have been overplayed, but it is nevertheless an important issue which emerges at several places in the book and deserves consideration.

The questions which form the core of this book clearly interrelate with each other closely; but they are not just about the Channel Tunnel. The Tunnel has been seen across Europe as symbolic in several different ways: as a symbol of the elimination of major barriers to transport and communication, as a symbol of closer ties, even as a symbol of the end of the threat of war. However, it is only one of a long list of major projects in Europe, from the crossing of the Øresund to the north to the Straits of Messina and Gibraltar in the south, and including Alpine and Pyrenean tunnels. The opening of Eastern Europe has demonstrated the enormous need for infrastructure renewal as a prerequisite of economic growth and integration. The pressure for new investment in infrastructure is equally great in the traditional core regions of Western Europe. In the context of plans for a new or upgraded 25 000 km railway network for Europe costing 150 billion ECU (£105 billion) also announced in December 1990, the Channel Tunnel seems a very small project. Since the intention and hope is that the private sector will be involved in these other projects, however, the lessons to be learned from the Channel Tunnel are enormous. Our research has already attracted considerable attention from around Europe, especially from those looking at the impacts of other major infrastructure projects.

Such a research project has involved a large team of people, bringing together expertise from different disciplines in law, economics and political science. The project has been co-ordinated in Lille by Gérard Marcou and in Canterbury by Roger Vickerman. Primary responsibility for legal questions has been assumed by Gérard Marcou, for economic questions by Roger Vickerman, and for political science questions by Ian Holliday. These three are responsible for the synthesis of material which appears in this book, which has been written by Ian Holliday (Chapters 1, 2, 4, 5, 6, 7, 8 and 11) and Roger Vickerman (Chapters 3, 9 and 10). Four other people have had a principal responsibility for basic research used in this book. On public policy and economic response in Kent and Nord-Pas de Calais: Michèle Breuillard and Michel Langrand. On questions of European integration and transfrontier co-operation: Clive Church and Yves Luchaire. In addition, direct inputs of work and valuable contributions to the development of the project in regular seminars (the proceedings of three of which — in December 1987, June 1988 and June 1989 — have been published as working papers) have been made by many others. In particular, we should like to acknowledge the contributions of Margaret Anderson, Martine Cliquennois-Lascombe, John Craven, Bernard Dolez, Serge Dormard, Gérard Duchêne, Richard de Friend, Bill Jenkins, John Kish, Christophe Masse, Charles Okeahialam, Chris Pickvance, and Alain Vantroys.

We have also been fortunate in the help we have received from many officials and elected members of local authorities and national government in both the UK and France, especially in Kent and Nord-Pas de Calais. In addition, we have had continuing dialogue with, and assistance from, Eurotunnel, British Rail, SNCF, Sealink, P&O European Ferries, and port and harbour authorities in Kent and Nord-Pas de Calais. It would be even more than usually invidious to single out individuals from any of these organisations who have given their time and energy to help our understanding. We are, however, particularly indebted to Andrew Lyall, who was responsible at the Department of Transport for work which led up to the decision of January 1986 and who, in retirement, has been of enormous help to our research.

The work included in this book has only been possible because of research funding. We are particularly grateful to ESRC (grant number YD00250018) and CNRS for their funding of this work under the auspices of their Anglo-French research programme. We are also grateful to the Région Nord-Pas de Calais, the French Ministry of Foreign Affairs, and the French Embassy in London for their support.

Production of this book has been greatly facilitated by the continuing good humour of Naomi Bothwell, the skill of Julie Beer in preparing maps, and the considerable patience of Iain Stevenson. As ever, final responsibility rests with the authors.

Ian Holliday
Roger Vickerman, Canterbury
January 1991

Gérard Marcou
Lille

Part One

The Channel Tunnel in context

1 *A brief history of the Channel Tunnel*[1]

Initially the romantic dream of a succession of (usually French) visionaries, the Channel fixed link matured quite quickly into a series of apparently viable projects, the most realistic of which was a set of tunnel schemes devised independently in the 1870s by teams of British and French engineers. However, more than a century was to pass between the elaboration of these detailed schemes and the start, in the 1980s, of serious and extended tunnel construction. A brief history of the Channel fixed link is therefore required to present and explain: the development of viable fixed-link options in the second half of the nineteenth century; the reasons for their non-realisation for a period of some one hundred years; the progress in the early 1980s towards a Channel fixed link once the earlier obstacles preventing construction had been removed; the selection of the Eurotunnel scheme in 1986; and the outlines of that scheme.

1.1 Early proposals: 1802–1882

The first known visionary was a Frenchman, Nicolas Desmaret, who in 1751 addressed a report, 'Dissertation sur l'ancienne jonction de l'Angleterre à la France' to King Louis XV. However, Desmaret's report was technically imprecise, and is consequently often overlooked by historians. By common consent, the first proper plan to construct a fixed link between Great Britain and France was a tunnel scheme for horse-drawn carriages, devised by Albert Mathieu-Favier (or Albert Mathieu), and presented to Napoleon in 1802 at the Peace of Amiens. Though primitive, Mathieu's submission, in contrast to Desmaret's, showed how two bored tunnels, linking Cap Gris Nez (near Calais) and Eastwear Bay (near Folkestone), with staging posts on the Varne Bank in mid-Channel, might be constructed. It is thus held to be the first in a long line of technically feasible (if not fully specified) schemes. The present Eurotunnel project is said to be twenty-seventh in this line.

Though favourably received, Mathieu's plan was never acted upon by Napoleon, due to other more pressing engagements. A rival immersed tube scheme, developed the next year, 1803, by another Frenchman, Tessier de Mottray, was also quickly forgotten. Nevertheless, the idea of a fixed link between Britain and France had been launched, and interest in it was to be dramatically revived in the middle of the nineteenth century by the railway boom, and by the engineering feats of great railway builders in both Britain and France.

However, it was not just the spur of commercial interest in railways which kept the project alive in mid-century, but also the dedication of one Frenchman, Thomé de Gamond, who from 1830 onwards conducted a number of often heroic studies and surveys in mid-Channel to plot the geological structure of the sea bed between the White Cliffs of Dover and Cap Blanc Nez in France. On the basis of this information, de Gamond developed a wide variety of schemes, ranging from bridges, viaducts and jetties to submerged and bored tunnels. He even proposed a transporter running on rails on the sea bed, but itself lifted clear of the sea.[2]

In 1857 the French government appointed a commission to consider de Gamond's bored tunnel scheme — reporting its findings in January 1858, the *Illustrated London News* forecast a six-hour rail service between London and Paris — but it was contact with a British mining engineer, William Low, which was to prove decisive to progress. Of de Gamond's many projects, Low saw a bored twin-rail-tunnel scheme in 1865, and on the basis of independent tests of the geological strata beneath the Channel, himself produced a similar proposal in 1867. This scheme was consistent with de Gamond's proposal, but conflicted with the single-bored twin-track scheme of Low's then British collaborator, John (later Sir John) Hawkshaw. The result was that a joint Anglo-French report in 1868, costing a bored twin-tunnel project at £8.5 million, carried the signatures of de Gamond, Low and another British engineer, James Brunlees, but was opposed by Hawkshaw. From this point, competitive projects on the British side were developed, and commercial interest was indeed the spur to progress.

To begin with, the lead was taken by Hawkshaw, who, with the backing of the London, Chatham & Dover Railway (LCDR), headed by James Staats Forbes, conducted a series of studies around St Margaret's Bay to the east of Dover, with the intention of developing a single-bore twin-track rail tunnel. Hawkshaw's English Channel Company, formed in 1872, gained statutory powers to acquire land and commence work through the British Channel Tunnel Company (Limited) Act 1875, which was paralleled by a French enactment granting concessionary powers to the Société Concessionaire des Chemins de Fer Sous-Marin, formed in 1874.

Low, meanwhile, sought financial support from the rival South Eastern Railway (SER), which approached the south coast via Tonbridge and Ashford (as opposed to Chatham), and was headed by Edward (later Sir Edward) Watkin, long-standing rival of Staats Forbes, and a man with sufficient control of railway companies to offer the prospect of circuitous, but nevertheless 'direct' rail services from Manchester to Paris. Watkin formed first, in 1875, the Anglo-French Submarine Railway Company, and subsequently, in 1881, the Submarine Continental Railway Company. On the basis of statutory powers granted in 1880, this company in 1881 started work on what was planned to be a bored twin-tunnel scheme at Shakespeare Cliff, to the west of Dover, and quickly began to rival Hawkshaw's workings in St Margaret's Bay to the east (see Figure 1.1).

At the start of the 1880s, then, two bored tunnel schemes were under competitive construction from the British side of the Channel, and a French tunnel was being driven from Sangatte. An Anglo-French Commission set up in the early 1870s had reported in favour of a Channel fixed link, and had helped to draw up two sets of national legislation, and a draft treaty to regulate matters such as international frontiers and the jurisdictions of national legal systems. Moreover, plans had been drawn up for this Commission to supervise not only construction work but also tunnel operation, charges and conditions of carriage, for the proposed 99 years of fixed-link concession. It seemed that the project which had first been seriously discussed at the start of the century would be realised before its close.

This, however, was not to be. All work on both Hawkshaw's and Low's schemes was first suspended in 1882, and then brought to a complete halt in 1883, by the British government. The primary reason given was national defence, an argument which continued to be presented by British governments until 1955. Then, on 16 February, in answer to a written question from the Labour backbencher, Lance Mallalieu, who asked 'to what extent strategical objections still prevent the construction of a road-rail tunnel under the Channel from England to France', Harold Macmillan, at the time Minister of Defence, answered 'Scarcely at all' (*Hansard*, vol. 537, col. 48). The defence argument had at last been officially undermined. However, it is now clear that in 1883 the argument had been more complex than this, and that it remained so until 1955 (and, indeed, has done so to the present day).

By the time of abandonment, both the Low tunnel at Shakespeare Cliff and the complementary French tunnel at Sangatte had reached a length of over a mile, and had defined the line which was to be followed by all subsequent tunnel schemes. Furthermore, a French geological commission, with which the hydrographer Larousse was associated, had produced a detailed geological map of the Strait which formed the basis for all subsequent studies, and had confirmed the view that if a submarine tunnel were driven between Dover and Sangatte, there would be every chance of finding a homogenous and regular chalk formation, suitable for tunnelling and sufficiently thick to contain a tunnel throughout its length. In subsequent corroboration of this view, the two tunnels, unlined, remained intact after abandonment, thereby giving practical evidence of the chalk's impermeability to water. The world's first compressed air rotary head tunnel-boring machine, invented by Colonel Beaumont of the Royal Engineers, remained in the tunnel at the foot of Shakespeare Cliff during a century of British opposition, only finally being removed when the present project was authorised.

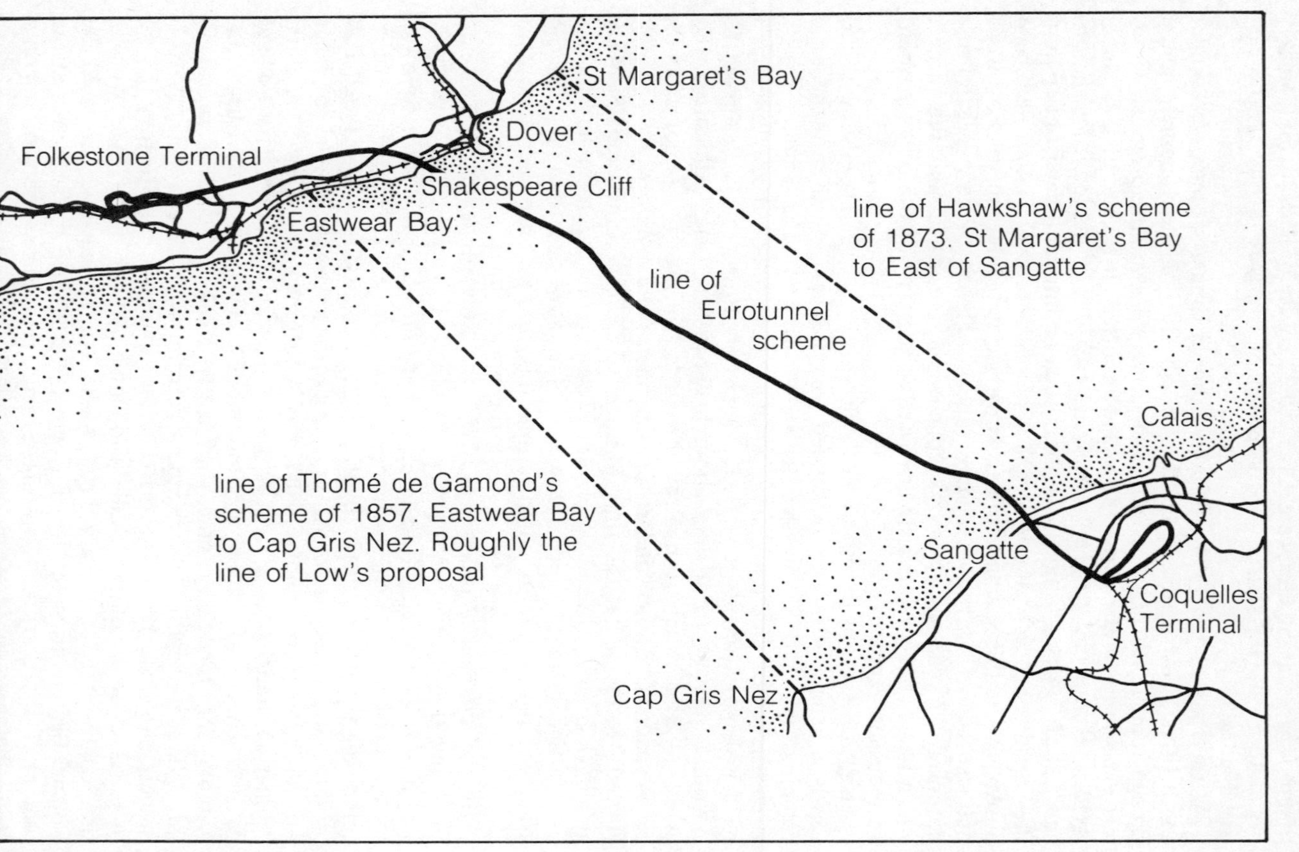

Figure 1.1 *Tunnel schemes*

1.2 A century of British opposition: 1882–1981

The single reason given by the British government for forcibly preventing Channel Tunnel construction from continuing in 1882 and 1883 was that of national security. Amid mounting unease in February 1882, a Channel Tunnel Defence Committee had been set up by the War Office to examine 'the practicability of closing effectually a submarine railway tunnel' in the event of war (quoted in Bonavia, 1987, p.34). Three months later, in May, it had reached a pessimistic conclusion, and had reported to the government that a Channel Tunnel was a potential danger to national security.

A Joint Select Committee of both Houses of Parliament, appointed as a direct consequence of this conclusion, had reaffirmed it in the summer of 1883, having been (unaccountably) impressed by a rambling and wildly inaccurate submission from Lieutenant-General Sir Garnet Wolseley, one of the highest-ranked army officers in the country.[3] Making both unfounded military objections to a tunnel, and sentimental appeals to a maintenance of Britain's island heritage, Wolseley had effectively sealed the case for the prosecution. On a split vote, authority to proceed with both tunnels was denied. Watkin, in 1890, abandoned his fight for the Channel Tunnel, and sought profit in the coal seam which had been found by tunnel work at Shakespeare Cliff. Ironically, it was discovered after his death that commercial mining was only viable in the territory of his great rival, Staats Forbes.

The defence argument was to remain the British government's reason for refusing authorisation of a Channel fixed link until the middle of the present century, when Macmillan made his 'Scarcely at all' statement in 1955. However, in many ways this argument was a screen, increasingly implausible, but useful nevertheless as a means of obscuring a complex set of British objections to a Channel fixed link, which went to the heart of the nation's identity, and which expressed a highly conservative notion of the country's proper links with continental Europe.

To begin with, the defence argument did not have to be articulated by government: a Channel fixed link was either technically impossible (at the start of the nineteenth century) or without realistic commercial potential (in mid-century). In these circumstances, visionaries and radicals could be allowed to scheme and plan. Sometimes they did so with clear ulterior motives. When, at the Peace of Amiens, Charles James Fox gave his support to Mathieu's scheme, it was clear that it was less a Channel Tunnel *per se*, more Napoleon and the ideals of the French Revolution (which Napoleon was then thought by some to embody), that Fox was promoting. Fox, it can be said, saw in the fixed link a means of transporting Continental notions of democracy to the shores of Britain's corrupt monarchy.

However, when it became clear, at the start of the 1880s, that a Channel Tunnel might actually be built, the British government was required to take sides. It chose to side with Lord Palmerston who, reacting in 1858 to British support for the de Gamond scheme considered by a

French governmental commission in 1857, had said: 'What! You pretend to ask us to contribute to a work the object of which is to shorten a distance we find already too short?' (quoted in Haining, 1989, p.6).

The government's chosen screen for this sort of reaction, forcefully expressed by Wolseley in his decisive memorandum to the Joint Select Committee of 1882–3, was never very plausible, and became ever less so with the passage of time and the development of modern warfare. Indeed, during and after the First World War, the defence argument was turned on its head, an operation which was to be repeated in the Second World War. Speaking at the end of the First World War, Marshal Foch maintained that a Channel Tunnel would have halved the war's length by allowing efficient despatch of the British Expeditionary Force and effective maintenance of its supply lines. In February 1913, in an article in the *Fortnightly Review* entitled 'Great Britain and the Next War', Sir Arthur Conan Doyle had already set out such an argument in favour of the Channel Tunnel, and it was to be passionately repeated by Churchill in the 1920s and 1930s. Each of these campaigns was to be in vain.

Despite the fact that, from the outset, the French had undermined the defence argument by suggesting that the approach at each end of a tunnel should have an exposed gallery which could be kept under the range of both shore-based and naval artillery, thereby guaranteeing the Tunnel's defence, the British military establishment held firm. Others were to note that in time of war a tunnel could very easily be flooded by the Navy. Nevertheless, well into the twentieth century the Imperial Defence Committee was to remain a useful resource for any politician seeking a 'respectable' reason to delay tunnel construction further: any reference to it was sure to be met with a clear warning of military dangers.[4]

Behind the thin screen of the defence argument, real opposition to a fixed link may be found. It tended to be based on sentimental attachment to Britain's island status — 'the reputation of England', Lord Randolph Churchill told Parliament in 1889, 'has hitherto depended on her being, as it were *virgo intacta*' (quoted in Bonavia, 1987, p.47) — and on xenophobia. Both sources of opposition to the tunnel remain strong today: Churchill's statement was echoed, curiously, by Barbara Castle, who wrote on cancellation of the 1970s project in 1975 of 'a kind of earthy feeling that an island is an island and should not be violated' (quoted in Bonavia, 1987, p.131); and in the mid-1980s Flexilink played on British xenophobia in mounting a campaign against a fixed link on behalf of port and ferry interests.

With the passage of time, however, opposition to the fixed link was to be considered increasingly reactionary, partly as the result of a tunnel's evident commercial interest. At the end of the nineteenth century, Gladstone had been an important champion of the Channel Tunnel, and during the period of the Entente Cordiale at the start of the present century, interest again vainly revived. In the period after the First World War, Churchill was to become equally fervent in the Tunnel's support (his father's fierce opposition notwithstanding). He wrote scathingly in 1924 of the 40-minute deliberations of 'five ex- or future ex-Prime Ministers' — Macdonald, Asquith, Lloyd George, Balfour and Baldwin — who decided, according to Churchill against the wishes of 400 Members of Parliament, to reject the Channel Tunnel on familiar grounds. 'There is no doubt about their promptitude', wrote Churchill. 'The question is, was their decision right or wrong? I do not hesitate to say it was wrong.'[5]

Right or wrong, it was upheld until 1955, when, at the very end of Churchill's premiership, the defence screen was lifted by Macmillan's House of Commons statement. After an interlude of three-quarters of a century, the project was again taken up with real possibility of progress. Further difficulties, again mostly British in origin, were, however, to delay the Channel Tunnel for a further quarter of a century: the present project was not launched until the start of the 1980s.

Part of the problem was that in the long period between official articulation of the defence argument in 1882–3, and official denial of it in 1955, there had been a substantial shift in transport use, both passenger and freight, from rail to road (and air). As a result, plans for a rail-only tunnel were no longer considered wholly acceptable by either the British or the French, and new ideas were sought. A French group, the Comité du Tunnel Mixte sous le Pas-de-Calais, had already, in 1938, put forward a preliminary plan for a combined road and rail tunnel, and, in the same year, had received backing from the French National Assembly. Once again, however, war had intervened.

The position after 1955 was therefore uncertain. The English and French Channel Tunnel companies, direct descendants of key protagonists of the 1870s and 1880s, remained in existence, but were now challenged by proponents of a road-rail tunnel with a drive-through option, developed after the war and publicly presented in 1958, and by supporters of a joint Anglo-French bridge project (again with a drive-through capability), recently established.

However, the traditional project — bored twin rail tunnels — quickly reasserted itself as front-runner. A Channel Tunnel Study Group, set up in July 1957 by the British and French Channel Tunnel companies in collaboration with the Suez Canal Company and Technical Studies Inc., an American firm, produced a substantial report in March 1960 which investigated all potentially viable fixed link options — bored tunnel, immersed tube, bridge, combined bridge and tunnel — and came out firmly in favour of a bored tunnel on both technical and financial grounds (Ministry of Transport, 1963). The work undertaken for this study was to form the basis for all subsequent reviews of fixed-link options. Its conclusion has never seriously been challenged, and its detailed tunnel plan has only been updated to reflect technical progress.

Thus, the final 20-year delay between the early 1960s and the early 1980s must be ascribed to lack of political will, compounded by adverse circumstances, both economic and political, at key points in the project's development. The publication in September 1963 of the report, *Proposals for a Fixed Channel Link* (Ministry of Transport, 1963), written by a collaborative team of British and French officials, and strongly supportive of a tunnel (as against a bridge) project, launched the 1960s scheme. In July 1966, plans were sufficiently advanced for the Prime Ministers of Britain and France, Wilson and Pompidou, to announce that a Channel Tunnel would definitely be built. In October of that same year, their responsible ministers, Castle and Pisani, were able to fix 1975 as the target date for its opening. Total construction cost (inclusive of terminals, rolling stock and ventilation) was put in the range of £155–170 million (1966 prices), as compared with the September 1963 estimate of £143 million (1962 prices). Most of the difference may be ascribed to inflation (Watson, 1967).

Further prevarication on the part of both governments, opposition from vested interests in Britain in particular — from the road lobby, port and shipping interests, part of the trade union movement, some local authorities in Kent, environmental groups, and so on — and the uncertainty in France caused by the May Events of 1968 rapidly made this target untenable. 'The record of this chapter [in the history of the fixed link] is depressing', wrote Bonavia (1987, p.98), 'since it seems to indicate the unfitness of governments to oversee a great public work.' Nevertheless, a project was launched in 1973, following publication of a more elaborate cost–benefit analysis (Department of the Environment, 1973b), which supported earlier decisions in favour of the bored twin-tunnel option. For the first time since the competitive tunnelling of Hawkshaw and Low in the 1880s, work preparatory to construction, including trial borings (covered by what was known as Agreement Number 1), was undertaken, and a second stage (Agreement Number 2) of full tunnelling activity was confidently awaited.

Total construction cost, which in the March 1973 Green Paper had been estimated at £366 million (1972 prices) (Department of the Environment, 1973a), was put at £468 million (1973 prices) in the White Paper published only six months later in September 1973 (Department of the Environment, 1973c). Each estimate contained a 10 per cent allowance for contingencies; most of the difference between them can be ascribed to inflation and exchange rate movements. Already, in these estimates, may be seen the cost inflation which was to contribute to abandonment of the 1970s project.

Cancellation, again by the British government, this time on 20 January 1975, was ascribed by Anthony Crosland, Secretary of State for the Environment, to two factors in particular: first, disruption to the Parliamentary timetable caused by general elections in February and October of the previous year, which had meant that the deadline of 1 January 1975 for ratification of the Channel Tunnel treaty (agreed with the French in 1973) could not be met; second, excessive increase in cost estimates for the high-speed rail link to the tunnel which the government was pledged to finance. The predicted cost of this had risen from some £120 million (1973 prices) in the 1973 White Paper (Department of the Environment, 1973c), to £373 million (May 1974 prices) when Crosland made a Parliamentary statement on the matter on 26 November 1974

(*Hansard*, vol. 882, col. 245).[6] Furthermore, as Crosland noted in his statement, the revised figure made no allowance for additional environmental work demanded in Kent and Surrey, for the changed compensation requirements introduced by passage of the Land Compensation Act 1973, or for the cost of a freight element to the link, which was now thought desirable.

However, the reasons for cancellation were again more complex than Crosland allows (Morris and Hough, 1986). It is true that the Parliamentary timetable had been disrupted, and that estimates of the cost of a dedicated rail link to the tunnel had risen sharply, but these need not have been insurmountable obstacles to the 1970s project. What was insurmountable was the obstacle of a weak (and in Kent very unpopular) Labour government, committed neither to the European Community nor to closer transport links with it, combined with escape clauses favourable both to private-sector investment in the tunnel (which effectively amounted to 10 per cent) and, paradoxically, to the British government, which was able to split all public-sector losses with the French.

At the moment of crisis, it quickly became apparent that no effective project champion existed on the British side, and abandonment became inevitable. Political will on the part of the Labour government could have averted cancellation, but its (understandable) lack is not solely to blame for the January 1975 announcement. Structural inadequacies in the framing of the Channel Tunnel's financing arrangements were also important contributory factors.

Six months later, a Channel Tunnel Advisory Group, established in April 1974 under the chairmanship of Sir Alec Cairncross by the minority Labour government which had taken office the previous month, reconfirmed the viability of a tunnel scheme (and provided perhaps the most independent and wide-ranging investigation of the Channel Tunnel ever undertaken at an official level in Britain) (Department of the Environment, 1975).

1.3 A decade of progress: the 1980s

Cancellation did not halt all tunnel-related activity, though it did bring systematic Anglo-French contact at governmental level to a close. The British Channel Tunnel Planning Council, set up in April 1971 under the chairmanship of a retired diplomat, Sir Eugene Melville, as the forerunner to the British half of a future Anglo-French Channel Tunnel Operating Authority, was wound up, and the project officially fell into abeyance.

Outside official Anglo-French circles, planning continued. A 1978 Coopers and Lybrand report, commissioned by DG VII, the transport directorate of the European Community, demonstrated the continued viability and thus commercial interest of a tunnel scheme, and prompted renewed interest in it (Commission of the European Communities, 1980). In April 1979, largely in response to this report, British Rail (BR) and its French equivalent, the Société Nationale des Chemins de fer Français (SNCF), produced a 'mousehole' single-track tunnel scheme (with accompanying smaller service tunnel) for rail passage only. 'Flights' of rail services in alternate directions through the tunnel were planned, with the London–Paris journey taking 4 hours 30 minutes, and London–Brussels 4 hours 10 minutes. No shuttle service for cars, lorries and coaches would be possible through the mousehole scheme's 6 metre diameter running tunnel. Estimated cost in January 1981 of this unambitious project was put by BR at £765 million, plus some £83 million for infrastructure investment at the British end.

This study was received without obvious interest by both the outgoing Labour Minister of Transport, William Rodgers, and his incoming Conservative successor, Norman Fowler. Indeed, the Department of Transport itself had not been seeking fixed-link projects in the late 1970s, and offered scant encouragement to BR. However, while there was little enthusiasm for a Channel Tunnel in 1979, the new Conservative administration was interested in exploring the possibility of private investment in transport. The mousehole scheme, bearing all the hallmarks of a public-sector scheme, did not itself fit the bill, but it did persuade Fowler, in a Commons statement in March 1980, to open the whole question out by inviting the private sector to submit schemes to him.

Initially, consortia such as the Cross-Channel Tunnel Group and Channel Tunnel Developments (1981) Ltd (a direct forebear of Eurotunnel, being sponsored by Tarmac and Wimpey) submitted plans to finance and build the BR–SNCF mousehole scheme, though neither was either detailed or robust. However, proposals quickly became more ambitious, and by late 1981 a whole host of schemes had been forwarded to the British government for assessment by a fixed-link team working in the Department of Transport. Also in direct response to Fowler's statement, the House of Commons Transport Select Committee decided to undertake an inquiry into 'the possibility of the construction of a fixed link for the transportation of passengers, vehicles and freight across the English Channel financed by private risk capital' (House of Commons, 1981, Vol. 1, p.vi). The French ministry, more suspicious than ever of the British since cancellation in 1975, and sceptical of the viability of private finance, did little or no fixed-link work at this time, and the project in the early 1980s looked like being purely British.

Progress was made, but not until the first Anglo-French summit to be held by Mrs Thatcher and President Mitterrand, in London in September 1981, was a political commitment to the principle of a fixed link given, and another Anglo-French study of viable options commissioned. The change in political will necessary to successful completion of a fixed-link scheme is traced by some commentators to this meeting, but at all levels relations between the British and the French remained cool with respect to the tunnel. The French, fully committed to a fixed link for almost every year of the previous century, were understandably suspicious of the British, being both mindful of the January 1975 cancellation (of which they had borne half the cost) and doubtful of the feasibility of private finance. The British, fully committed to private funding, were sceptical of the (then radical) Mitterrand government's likely acquiescence in it.

Nevertheless, independent exploratory studies were undertaken in both countries, and the collaborative group of officials working under Andrew Lyall, Under Secretary in the British Department of Transport, and Guy Braibant, Chargé de Mission in the French Ministry of Transport, continued its assessment of fixed-link options. In Britain, studies were most concerned to evaluate the feasibility of private funding for a fixed link, and were conducted mainly by National Westminster Bank and Midland Bank, in collaboration with Crédit Lyonnais. The Banking Group's report, issued in May 1984, had two strands, one British, one French. Each concluded that a rail tunnel was the only viable option, and that government financial guarantees were imperative (Franco-British Channel Link Financing Group, 1984). In France, studies were most concerned to assess opinion in the region directly affected by fixed-link construction and operation, Nord-Pas de Calais. A series of meetings, both public and private, held in key parts of the region by Braibant at the end of 1981 concluded that on balance opinion was strongly favourable. Pierre Mauroy, Prime Minister of France and Mayor of Lille, was very keen, as was his Transport Minister, Charles Fiterman (who had no power base in the North).

In Britain, little equivalent enthusiasm in high places could be detected, though David Howell, Fowler's successor as Transport Minister, was a fixed-link advocate. At Cabinet during the Falklands War, proposals for a fixed link received a lukewarm response from Mrs Thatcher, who was, however, preoccupied with military matters: the fixed link happened to be discussed on the day of the sinking of HMS *Sheffield*.

Real political will on the British side seems to have been generated three years after the first Thatcher–Mitterrand summit, at the end of 1984, when Mrs Thatcher, in the British embassy in Paris, decided that it was time to do 'something exciting'. To British officials involved in the project from the start of the 1980s, the Prime Minister's enthusiasm for the fixed link apparently emerged *at this point* (and no sooner). Such is the nature of Britain's constitutional arrangements, and such was their particular configuration under Mrs Thatcher, that real political will was generated with this shift in the Prime Minister's attitude.

The shift itself may be explained by the nature of Britain's then relations with France and the rest of the EC. Only at the Fontainebleau summit, in June 1984, had the issue of Britain's contribution to the EC budget been finally resolved. With this issue behind her, notes Hugo Young, 'Mrs Thatcher began to utter sentiments that were impeccably *communautaire*', among which the new line in favour of the fixed link must be counted (Young, 1990, p.388). Strangely, in view of the British government's subsequent policy, Mrs Thatcher even echoed Ernest Bevin's description of the aims of British foreign policy when addressing the Franco-British Council in Avignon on 30 November 1984: 'To go down to Victoria Station, get a railway ticket and go where the hell I like without a passport or anything else' (ibid.). The present Channel Tunnel project would thus seem to have been launched by a rare period of Euro-enthusiasm on the part of the British Prime Minister.

Coinciding as it did with considerable French interest in the fixed link, both nationally and regionally, political drive from the top of the British government was now able to give the project substantial momentum. A three-month timetable was set for the Anglo-French Study Group of officials to draft an *Invitation to Promoters* (of potential fixed links). This was issued jointly by the British and French governments on 2 April 1985, with a deadline for submissions of 31 October 1985. A further two months were then set aside for evaluation of submissions by the Study Group, allowing announcement of a private-sector fixed-link concessionaire to be made at the beginning of 1986.

Only by keeping to this tight timetable would the two governments be able to make an announcement before the French parliamentary elections scheduled for March 1986 (which were very likely to produce a change of government, and a National Assembly hostile to the President). In the event, the choice was announced at Lille town hall on 20 January 1986. Two months later, the incoming Chirac government found itself in possession of a Concession Agreement and a Treaty negotiated by its Socialist predecessor.

After more than a century of (usually British) problems, ranging from prevarication to (more often) outright opposition, the two governments had concluded a Concession Agreement with the Channel Tunnel Group Limited and France-Manche SA, and a treaty with each other, in the space of little more than twelve months. The concessionaires quickly reformed themselves as Eurotunnel, and, with the sponsorship of the two governments, set about gaining parliamentary consent for their approved scheme. In France, this was a straightforward matter, conducted through the normal public inquiry system. In Britain, the device of a Hybrid Bill was used, and took up large amounts of Parliamentary time. Royal Assent to the Channel Tunnel Act 1987, given on 23 July, triggered ratification of the treaty six days later, and formal authorisation of the Eurotunnel project.

1.4 Selection of Eurotunnel

The 1985 competition among fixed-link proponents laid down four basic rules: entries must be technically feasible, financially viable, Anglo-French, and accompanied by an Environmental Impact Assessment (EIA) (undertaken, for the first time in the Community, in accordance with the recent EC Directive 85/337 on EIAs). Entrants were also required to deposit a sum of 300 000 ECUs with the two governments.

Ten entries were received by the deadline of 31 October, some of which breached some or all of the rules set out in the *Invitation to Promoters*. Four contenders were thought to be serious, though even among this group adherence to the competition rules was variable. These contenders, which were submitted to detailed examination by government assessors, were:

(1) Channel Tunnel Group/France-Manche (later to become Eurotunnel): twin bored large, 7.3 metre (later 7.6 metre) diameter, tunnels carrying shuttle and through trains;
(2) EuroRoute: bridge/tunnel scheme comprising bridges from each coast linked by a submerged tube tunnel 21 km long carrying a motorway, plus a bored small-diameter tunnel system for through trains, which would be built in stages;
(3) Eurobridge: bridge scheme comprising a motorway in an enclosed tube suspended from piers in spans 4.5 km in length, using advanced plastics technology; a rail link could be provided either on the bridge, or in a small-diameter tunnel;
(4) Channel Expressway: twin very large bored tunnels, 11.3 metres in diameter, carrying both motorway and rail traffic (later modified to comprise separate rail tunnels), using advanced ionisation techniques to remove noxious gases from road vehicles.

The members of the four consortia are listed in Table 1.1.

The process by which the CTG-FM scheme was selected was mysterious even to those closely involved in it. The British government in particular was concerned not to publish assessments of financial viability so as not to interfere with the workings of the market. However, official secrecy extended further than this. Sir Nicholas Henderson, at the time chairman of CTG, confesses himself unable to understand, or even to follow, the various manoeuvres made by other contenders, reacting to perceived ministerial preferences, in the period between 31 October 1985 and 20 January 1986 (Henderson, 1987). It seems that British government officials were equally baffled by the pattern of shifting proposals and alliances made by fixed link contractors at this time.

Assessors evaluated the four serious candidates in terms of technical viability, financial robustness and environmental impact. The House of Commons Transport Committee also examined the schemes (House of Commons, 1985). In Britain, evaluation took place against the background of earlier preferences for a drive-through scheme, ultimately traceable, it is said, to 10 Downing Street, and the Prime Minister's antagonism towards railways and rail unions. In France there were signs that traditional preferences for a rail-only project, were generating support for the road-rail EuroRoute scheme. This scheme offered both a large French presence and the prospect of large numbers of jobs, not just in depressed Nord-Pas de Calais, but also on fabrication work in other areas of France. Enthusiasm for this option was said to come from the Elysée.

The early favourite was EuroRoute, which seemed to many observers to be both technically feasible, and desirable to all parties. However, British assessors questioned the viability of this scheme on both technical and financial grounds.

For a long time, therefore, it appeared that when the British side spoke

Table 1.1 *Major scheme sponsors, 1985*

UK	France
Eurotunnel	
Channel Tunnel Group	*France-Manche*
Banks	
National Westminster Bank	Banque Nationale de Paris
Midland Bank	Crédit Lyonnais
	Banque Indosuez
Contractors	
Balfour Beatty Construction	Bouygus
Costain UK	Dumez
Tarmac Construction	SPIE/Batignolles
Taylor Woodrow Construction	Soc. Auxiliaire d'Entreprises
Geo. Wimpey International	Soc. Générale d'Entreprises
EuroRoute	
EuroRoute	*EuroRoute France*
Concession companies	
Associated British Ports	Alsthom
Barclays Bank	Banque Paribas
British Steel Corporation	CGE
British Telecom	GTM Entreprises
Kleinwort Benson	Société Générale
Trafalgar House	Usinor
Construction companies	
EuroRoute Construction	*Scoltram*
British Shipbuilders	Alsthom
British Steel Corporation	GTM Entreprises
GEC	Usinor
John Howard	
Trafalgar House	
Eurobridge	
Banks	
Arbuthnot Latham	
Contractors	
Imperial Chemical Industries	
Dupont Fibres	
Brown & Root	
John Laing	
Channel Expressway	
Sea Containers Ltd	
Banks	
	Crédit du Nord
Contractors	
	SCREC

of drive-through schemes, it had in mind Channel Expressway, the road-rail scheme entered at the very last minute by Sea Containers, the parent company of Sealink British Ferries, which until then had been a member of the Flexilink consortium which opposed all fixed-link schemes and promoted instead the 'flexible' alternative provided by ferries. It, too, took up the running at one stage. However, this scheme had unresolved technical difficulties. Initially, it was planned to run road and rail traffic at intervals in the same tunnels, a bizarre proposal which was readily altered by the simple expedient of separating the two types of traffic into dedicated tunnels. However, very real problems of exhaust ventilation from cars, coaches and lorries remained, and were never satisfactorily resolved by Channel Expressway. Furthermore, its almost total lack of French backers meant that it was not held in high esteem by the French government. Though very much in favour in Britain at one stage, James Sherwood, the American chairman of Sealink, was very much out of favour by the time the decision was taken. No clear explanation for either the original position or the end state has ever emerged.

Finally, frequent attempts were made by both governments — and particularly by British Transport Minister, Nicholas Ridley — to unite the various competitors around a single project, in order to maximise financial, technical and managerial support for the fixed link (Henderson, 1987). However, no progress of this kind was ever made, and it seems in many ways absurd that Ridley should have expected serious contenders like CTG and EuroRoute to have anything to do with Sherwood. Granted he clearly had influential friends on the UK side (he happened to be Ridley's neighbour), but his scheme was unrealistic, and, as a ferry operator, he was in any case mistrusted by the other contenders.

Eventually, at Lille town hall on 20 January 1986, Mrs Thatcher and President Mitterrand announced their selection of CTG and FM. The next month, in Canterbury, Britain and France signed a draft treaty (which was ratified on 29 July 1987, following the successful passage of Channel Tunnel legislation in both Britain and France), and the month after that, in Paris, the two governments signed a concession agreement with CTG and FM. The way was now open for two sets of national legislation, authorising the project, to begin their passages through the British Houses of Parliament and the French National Assembly. This passage was accomplished rather more quickly in France than in Britain.

In many ways the selection of CTG and FM was a compromise solution, and both leaders are said to have been unhappy with the choice made. However, it was relatively safe, in that CTG-FM's Tunnel scheme depended on proven technology, looked financially viable, and was a clear extension of projects which had been positively vetted by official commissions in the 1960s and 1970s. The concession agreement signed with CTG and FM, which provides for a concessionary period of 55 years (from the date of treaty ratification, 29 July 1987), also holds open the possibility of future drive-through schemes, through clauses which state that unless CTG-FM devises a drive-through option by 2010 its monopoly of fixed link concession can be withdrawn after 2020.

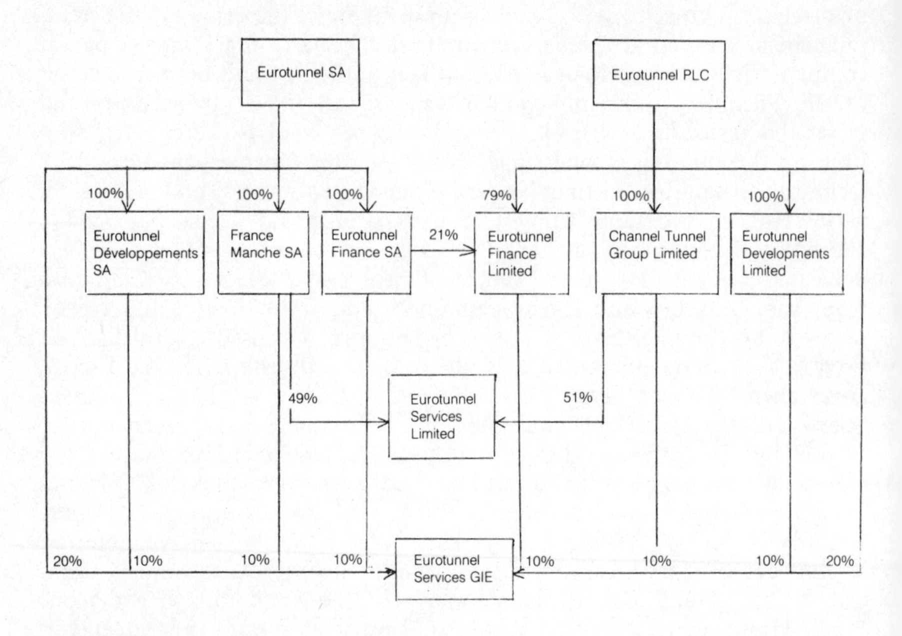

Figure 1.2 *Eurotunnel corporate structure*

Source: Eurotunnel (1990b)

1.5 The Eurotunnel scheme

The victorious concessionaires were both consortia of leading construction companies and banks in each of the two countries (see Table 1.1). They quickly formed two holding companies, Eurotunnel PLC and Eurotunnel SA (see Figure 1.2), which were given the job of raising finance, and an umbrella holding company, the Eurotunnel Group.

Thus, the CTG-FM scheme became the Eurotunnel scheme. It comprises twin 7.6 metre (initially 7.3 metre) running tunnels, 50 km in length (of which 38 km is under the Channel), linking Cheriton near Folkestone and Fréthun near Calais. In addition, there is a 4.8 metre diameter service tunnel with linking passages to the main running tunnels at 300 m intervals, plus two cross-over points between the running tunnels at one-third distances along the tunnel from each end. At each terminal, the Tunnel system is linked to both national motorway and rail systems.

The concession agreement provides for certain minimum standards of service in terms of maximum headways in off-peak periods, and maximum delays in the busiest periods. The concession agreement and the Channel Tunnel Act 1987 also provide some regulation of the competitive environment by specifying the commonality of any frontier restrictions, availability of duty-free concessions, and so on, which should apply to both ferries and the Tunnel. Section 42 of the British

Table 1.2 *Geographical breakdown of Eurotunnel loans*

	Nov 1987 (%)	Nov 1990 (%)	Total (%)
France	18	25	20
UK	10	19	12
Germany	13	13	13
Other Europe	24	17	22
Total Europe	65	74	67
Japan	23	22	23
USA	2	3	2
Rest of world	10	1	8
Total non-Europe	35	26	33
World total (£m)	5000	1800	6800

Act specifies that no financial subsidy can be made by government to through rail services (passenger and freight) using the Tunnel.

To finance the Tunnel, Eurotunnel sought both equity and loan capital, the latter being to some extent conditional on the former. The initial equity interest of founder shareholders in Eurotunnel became known as Equity I. In October 1986, a private placement in Equity II was arranged (partly through the good offices of the Bank of England on the UK side). In November 1987, hot on the heels of the October stock market crash, the main public flotation in Equity III was launched. Again, difficulties were experienced in the UK, and the issue was eventually fully underwritten, though undersubscribed by the public.

Each equity share in Eurotunnel is an indivisible unit comprising a share in Eurotunnel PLC and a share in Eurotunnel SA. Loan finance, in the initial form of a syndicated loan of £5 billion, has been raised through a consortium of around 200 banks world-wide (of which very few are British). The geographical breakdown of Eurotunnel's loans is presented in Table 1.2. £1 billion of this £5 billion loan comes from the European Investment Bank (EIB), unusually guaranteed in this case not by the recipient states' governments, but by the consortium of commercial banks. In view of subsequent cost overruns, both equity and loan capital had to be increased beyond the combined £6 billion which had been raised by the end of 1987. The increase in equity took the form of a rights issue in November 1990, which raised a further £566 million. Simultaneously, additional loan capital to the tune of £2.1 billion was raised, £1.8 billion coming from the banking syndicate, and the remaining £300 million being a parallel line loan from the EIB. The unusual, indeed unique, feature of this loan is that it is not secured by letters of credit, but on Eurotunnel's existing assets.

To build the Tunnel, Eurotunnel contracted Transmanche Link (TML), thereby generating a proper client–contractor relationship at the

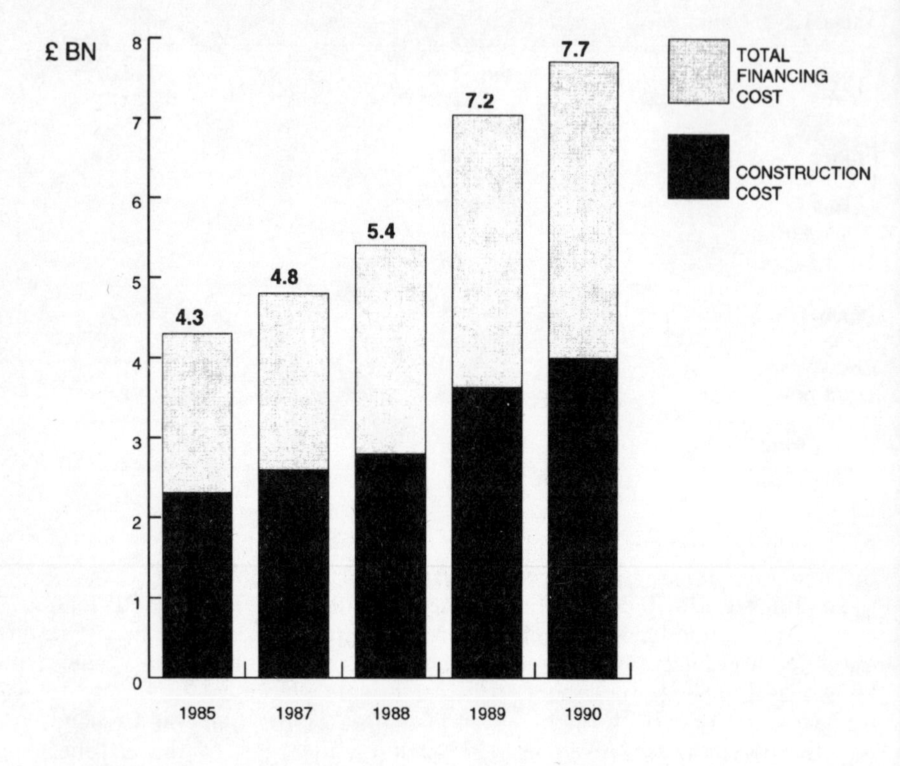

Figure 1.3 *Construction cost increases*

heart of the project. TML is an Anglo-French joint venture between Translink in the UK, and GIE Transmanche Construction in France, these two groups in turn being joint ventures of the construction companies originally brought together in CTG and FM.

To monitor Eurotunnel's progress and operations, the loan syndicate has appointed consultants, and the two governments have appointed an independent *maître d'oeuvre*. The *maître d'oeuvre* reports to an Intergovernmental Commission of British and French officials, which is the main overseeing body created by the Tunnel Treaty and Concession Agreement. It has subsidiary commissions, of which the Safety Authority is the best known. Eurotunnel's operations and proposals are thus subject to considerable external control.

Initially 15 May 1993 was set as the Tunnel's opening date, but early construction delays — or, as TML maintains, delays in starting the project up — caused this to be put back by a month, in April 1989, to 15 June 1993. Meanwhile, real construction cost estimates rose (in 1985 prices throughout) from around £2.3 billion in the 1985 submission, to £2.7 billion at the time of the Equity III prospectus in November 1987, to nearly £3 billion a year later. At the end of 1989, they were estimated by Eurotunnel at £4 billion, by TML at £4.2 billion, and by the banks' technical adviser at a possible £4.6 billion. Total financing costs, including allowance for inflation, which had been estimated at around £4.8 billion in 1987, and at £5.4 billion in 1988, were thought at this time to be anything between £7.5 and £8 billion (see Figure 1.3).

As a result, in January 1990, crisis point was reached. By suggesting that even TML's latest figure of £8 billion for cost to completion was too low, the banks' independent technical adviser put the entire project in doubt. The sense of crisis which hung over the project at this time was amplified by a number of personality clashes, notably involving Alastair Morton, from February 1987 until February 1990 co-chairman of Eurotunnel with André Bénard. In a series of senior management changes made in February 1990, as part of a revised contractual agreement with TML, Morton was made chief executive under Bénard, who became sole chairman. These changes were designed to restore (bankers') confidence in a project which, within the space of only two or three years, had come to be perceived as dangerously close to terminal difficulties.

Following agreement with TML, Eurotunnel put total construction cost at £4.2 billion, and total financing cost to completion at £7.6 billion (although some elements had been referred to arbitration). To maintain the desired funding margin, Eurotunnel was required to raise an extra £2.7 billion, of which £2.1 billion would come from loans, and the remainder from new equity through a rights issue, as was mentioned above.[7] The refinancing operation was concluded at the end of 1990 with a successful rights issue, and the project now looks both financially secure and, with breakthrough in the service tunnel on 1 December 1990, technically assured.

The remainder of this book will explore some of the issues raised by the decision taken in the 1980s to build a Channel Tunnel.

Notes

1. This chapter draws heavily on Bonavia (1987), and uses material from House of Commons (1986, Vol. 1) and Haining (1989).
2. The idea is not totally unworkable. At Saint-Denis-sur-Seine, in 1896, a M. Ernest Bazin actually built a 'Roller-Boat' for 'stable' Channel crossings. Some 125 m long, and indeed lifted clear of the waters of the Seine, the craft was propelled on rails by a steam-driven screw running between huge rollers. It never went into commercial operation.
3. Wolseley's memorandum is reproduced in Haining (1989, pp.73–5, 78–90).
4. Thus, the switching of a few votes in the Parliamentary division of June 1930, which went 179–172 against resumption of tunnelling activity, would probably not have made any difference: the Committee had already been primed to advance its usual objections.
5. Churchill's article in the *Weekly Dispatch* of July 1924 is reproduced in Haining (1989, pp.110–11).
6. A measure of the project's inflationary problems may be gauged from the shift in the price index from an annual to a monthly base.

7. The financial basis of the Channel Tunnel project is discussed in greater detail in Chapter 3. See specifically Table 3.1 for a detailed financial summary of the project.

2 *Planning and authorisation*

Writing at the end of the 1970s, Peter Hall (1980) listed the Channel Tunnel among his *Great Planning Disasters*. It was, he said, a negative disaster — like London's third airport — where considerable resources of both time and money had been invested in a scheme which was later abandoned (as opposed to a positive disaster — like Concorde — where considerable resources are poured into a scheme which many informed people recognise to be a mistake). The purpose of this chapter is to investigate the processes used to plan Channel Tunnel construction and operation in Britain and France, both (briefly) in the 1960s and 1970s, and (at greater length) in the 1980s. In seeking to determine why the 1970s scheme was a negative planning disaster, and how the 1980s scheme has sought to resolve its inherent problems, this chapter contributes to a major theme of this book, which is consideration of whether the Tunnel can in any sense be termed a positive planning disaster (the assumption being that it will be completed).

2.1 The 1970s scheme

The 1970s scheme, which began life at the start of the 1960s, but which was not properly launched until 1973, was ostensibly abandoned by the British government in 1975 for two reasons, mentioned in Chapter 1: electoral disruption of Parliament, and cost overrun (on the high-speed rail link from Folkestone to London). However, as has already been noted, lack of political will on the part of a weak Labour government facing internal dissension (over Britain's relations with the rest of Europe) and external protest (over the impact of a high-speed rail link on Kent), combined with escape clauses which gave private-sector interests greater incentive to acquiesce in the government's decision than to fight it, were more decisive factors.

In these circumstances, the disastrous element of the regime set up to plan the Tunnel in the 1970s would seem to be the lack of a project champion at the centre of it. As Morris and Hough (1986, p.52) note: 'The tragedy of this project — perhaps the real reason for its failure — was that it did not have an owner'. A crucial counterpart to this was that its costs were to a large extent contractor-driven.

When costs duly spiralled, the project was allowed to drift into crisis, and no one came forward to press the case of the Tunnel itself in discussions of what was to be done. For many reasons, the Labour government, direct heir of the 1964–70 Wilson government which had enthusiastically promoted the fixed-link option in 1966, found it a great deal easier to abandon the Tunnel than to press ahead with it. For other, equally good reasons, private-sector interest in the Tunnel found it easier to take the money and run than to continue to support a project which had an uncertain political future. In this situation, there was little the aggrieved French government — perhaps the sole remaining enthusiast for the scheme in a position to do anything about it — could do but pay its half of the cost of cancellation.

It is almost certainly overstating the case to conclude, as Morris and Hough do, that but for an unfortunate mix of circumstances, the 1970s scheme would have been wholly successful. 'Had the Labour Government not been elected,' they write, 'the Tunnel would probably have been built successfully, in time, in budget (bar exchange rate variations), at a profit, to just about everyone's satisfaction' (Morris and Hough, 1986, p.55). On the one hand, it is quite possible that a Conservative government re-elected under Mr Heath in 1974 would have come to the same decision as did its Labour successor in 1975, for, as has been shown, the crucial defect of the 1970s scheme was structural, not merely political: in hard times, economically and politically (which 1974 and 1975 would have been for any British government), the project offered tempting escape clauses to almost all parties. On the other hand, it is by no means certain that if the 1970s scheme had survived crisis in 1974–5, it would thereafter have had a trouble-free existence: the structural problems which in fact killed the project in 1975 would have remained as substantial impediments to successful completion in subsequent years (as has been shown to some extent by the experience of the 1980s scheme).

Thus, it is possible to argue that the 1970s scheme had the potential to be either a negative great planning disaster (which in the event it was), or a positive one. Its crucial structural defects meant that it was guaranteed to 'fail', whatever happened. It is in this context, and on this basis, that the 1980s scheme must be assessed.

2.2 The legacy of the 1970s scheme

In Britain, clear lessons were learnt from the 1970s scheme: that, as Bonavia (1987) dolefully notes, governments are not particularly good managers of major projects, that the politics of the Tunnel, and of a high-speed rail link to it, is a potentially explosive issue in Great Britain, and that the electoral cycle and Parliamentary procedures need to be carefully considered when dealing with a project of the magnitude of the Tunnel.

Each of these lessons was learnt in a particular way by the Conservative administrations of Mrs Thatcher elected after 1979. Fully aware not only of the experience of the Channel Tunnel in the 1960s and 1970s but also of other large public projects, particularly Concorde and the proposed third London airport at Maplin Sands, the government considered itself unfit not only to manage similar future schemes but also in large measure to plan them. For, what the experience of the 1960s and 1970s was said to demonstrate was that the market is both the best test of need and the most efficient discipline on management.

It is important to stress that this lesson was drawn not just from experience of the Channel Tunnel in the 1960s and 1970s. Concorde and Maplin were the two other clear instances of great planning disaster from that period, but there were also enough examples of lesser planning disasters to create a climate in which the Conservative government could both retreat from planning, and find that its policy was approved by important sections of the population. It is also important to note the specific lessons which the government drew from Concorde and Maplin.

From Concorde it learned that successive governments had been locked into an increasingly costly contract with virtually no means of escape. In his memoirs, Harold Wilson claims that in 1964 the incoming Labour government sought to cancel the Concorde project, but found itself unable to do so (Wilson, 1971, pp.61–2). Interestingly, Bruce-Gardyne and Lawson (1976) corroborate Wilson's account, and draw some of the lessons from Concorde listed here. Each was influential in Mrs Thatcher's Conservative Party, both in opposition and in government.

From Maplin it learned the folly of high-profile governmental identification with, and promotion of, a major project with significant environmental implications. To many observers, Maplin was an undesirable Tory project, and it was partly for this reason that it eventually caused the managerial Heath government enormous difficulties. It was quickly cancelled by Labour on its return to power in 1974.

The politics of the Tunnel in the 1960s and 1970s (as opposed to government administration of it) taught that government is advised to shift on to other bodies, preferably outside the public sector, responsibility for as many aspects as possible of a project as controversial as the Tunnel. A policy of political withdrawal, it was argued, would have a dual benefit: it would both relieve pressure on government, which was good in itself, and, by eliminating an actor which was commonly thought able to resolve any problem through an injection of public funds, would make a 'realistic' resolution of difficulties more likely.[1] Each argument was widely applied in the 1980s: to investment decisions in industry, and to public-sector pay bargaining, for example. Furthermore, a policy of political withdrawal was highly consistent with the earlier lesson concerning public management of major projects.

The third lesson, that Parliamentary business and the electoral cycle need to be carefully managed, concerns the residual role of government in the Conservative vision of major project management. In this area, as in the other two, the government's strategy was at times controversial, for it comprised an investigative procedure which was seen by some as politically manipulated to fit an electoral cycle which could certainly delay the 1980s project, and might even wreck it. At the centre of this procedure was a hybrid Parliamentary bill which obviated the need for a public inquiry on the British side, and was thus alleged to be both a departure from normal practice and a denial of important democratic rights of participation and objection.

The lessons the French drew from the 1970s scheme were different from those drawn by the British. Chiefly, they were that the British could not be trusted, especially when it came to discussing a possible Tunnel between Britain and France, and that, therefore, cast-iron political guarantees should be written into any future fixed-link treaty. One lesson which the Socialist administration of the early Mitterrand years did not draw from the experience of the 1970s was that the private sector should henceforth be entrusted with Tunnel planning and management. From a French perspective, the private sector had proved itself as irresponsible in 1975, as had the British government, and successful realisation of a fixed-link scheme required clear public-sector direction.

Thus, when fixed-link discussions were reopened at ministerial and official levels at the start of the 1980s, British and French negotiators had fundamentally different ideas about the role of government in a future fixed-link scheme. Initially, these different ideas threatened yet again to prevent progress. Indeed, French negotiators for some time believed that the British government was promoting an implausible arrangement — to concede construction and operation of a fixed link to private enterprise — merely as a means of sabotaging a project which had never been particularly popular in the UK, and which remained unpopular at the start of the 1980s. However, gradually it became clear both that this time the British were negotiating in good faith, and that they were fully committed to 'privatisation' of the fixed link. Indeed, it was eventually impressed upon the French that no other arrangement would be considered by British negotiators, and that progress would be possible only on these terms.

2.3 The 1980s scheme

The status envisaged, and devised, for a fixed link by British negotiators was that of a government franchise, or concession — initially, the two terms will be used interchangeably — to the private sector. Thus, in 1985 the two states held what was in effect a fixed-link competition, which was judged by their officials, and won by CTG-FM (later Eurotunnel). At this point, machinery in each country was invoked to concede Channel Tunnel construction and operation to the victorious consortium, which was then to assume sole responsibility for both operations (until the concession was terminated, after a period of 55 years, and Tunnel operation handed over to the two states).

The role of the two states in the 1980s project was thereby diminished (in comparison with the 1970s project), but it was not reduced to a minimum, either in possible or in historical terms. As in the 1870s and

1880s, when there was competitive tunnelling on the British side of the Channel, it would have been perfectly possible to reduce the two states' involvement to that of signing an accord authorising construction of a fixed link between Britain and France, and of allowing the market to determine whether and how it should be built. This would have been the role of the minimal state. In fact, the role of government in the 1980s project has been a great deal more substantial than this, though it has not been promotional in the manner of the 1970s project.

In essence, the role of government in the 1980s has been regulatory, in contrast to the minimal role of the 1870s and 1880s, and to the (literally) constructive role of the 1970s. The two states set fairly precise rules by which the 1985 competition was run, and they have since placed a number of restrictions on Eurotunnel's commercial freedom. In so doing, they have developed a form of concession which is pure in theory, though sometimes corrupted in practice (particularly in France). In a discussion of the planning regime within which the Tunnel is set, it is important to understand the nature of Eurotunnel's concession. This can be done by means of a dual contrast: with the tradition of state concession which has been developed in France over the course of many decades; and with the type of commercial franchise which has been developed in the USA, and subsequently in most parts of the world, in recent years.

The French concessionary tradition will be discussed first, for it is the experience French officials brought to negotiations with their British counterparts. Against this, the Channel Fixed Link Concession Agreement, developed by British and French officials, will be set. It diverges from recent French practice but, as will next be demonstrated, does not fully conform to the American-style commercial franchise.

French concession in theory and practice

The French concessionary tradition is long and highly developed; it has no British equivalent. It can certainly be traced to the railway boom of the mid-nineteenth century, when expansion of the French network was accomplished chiefly by means of a series of concessions, and it has since been a major feature of French life. At the end of the last century and the beginning of this, most major services — rail, water, gas, electricity — were operated by concession. In the period since the Second World War, the legal device of a concession has been used with increasing frequency: to extend the French motorway system, to build road-bridges and tunnels, to develop residential and industrial zones, to exploit (small) ports and marinas, to augment hospital provision, and to improve municipal facilities. Previously, each of these areas had been reserved for public provision (Dufau, 1979).

The classic French definition of concession states that it is a contract by means of which a public person authorises a private person to provide a public service *à ses risques et périls*, the private person being remunerated by service users (Dufau, 1979, p.3). In principle, then, a French concessionary contract is uncomplicated: mere authorisation is provided by the public authority (whether national, regional or local), and full responsibility for service finance and delivery is assumed by the concessionaire. In practice, French concessions have turned out to be highly complex, as concessionaires have found themselves in financial difficulty, and have turned to the public authority for assistance. This practice is well illustrated by the history of French motorway concessions.

When, by a law dated 18 April 1955, one of the many governments of the Fourth Republic for the first time authorised motorway concession, it maintained that this practice was 'irregular' or 'exceptional', and held both that concessionaires must have a majority public-sector interest, and that use of motorways should remain free. This, however, proved to be a costly stipulation for the public sector, and threatened to undermine the ambitious road-building programme on which France was embarked. Indeed, there was little point in extending the concessionary device to the construction and operation of motorways if it was unable effectively to tap private capital. Thus, by a series of laws in the 1960s and early 1970s, successive governments introduced the twin concessionary principles of private-sector provision, and payment by service users, to motorway practice.

In one sense, then, French motorway concessions approached the classic understanding of concession. However, in another sense they departed in significant ways from it. On the one hand, the state offered financial guarantees to all motorway concessionaires. On the other, it regulated the tolls which they could charge (Touret, 1972). Such was the 'crisis' in motorway concession (Zoller, 1979) that by the beginning of the 1980s, conceded French motorways were almost universally in receipt of public assistance, the sole exceptions being the Paris–Poitiers and Paris–Rennes routes, both of which were managed on a concessionary basis by the Compagnie Finanacière et Industrielle des Autoroutes (COFIROUTE) (Fayard, 1980).

Intimately connected with the history of French motorway concessions are increasingly developed links between the public and private sectors throughout the economy, particularly (again) in the post-war era. In a report written in 1960–1, the Cour des Comptes was already able to refer to 'le démembrement de l'Etat', and 'le démembrement des services publics', when discussing the recent evolution of French public administration (Touret, 1972, p.376). By this, it meant that an array of intermediate bodies had latterly emerged in France both to take the place of traditional forms of public authority and to undertake traditional public service tasks. These *sociétés nationales*, *établissements publics* and *sociétés d'économie mixte* (SEMs) now have an extensive presence in the French economy. At the national level, *établissements publics* are represented by such companies as SNCF, Air France, Electricité de France and a number of shipping companies. At regional and local levels, SEMs are prevalent, motorway concessionaires being good examples at the regional and pan-regional level.

In the conceded motorway sector, the importance of SEMs has significantly increased in the 1980s. Of ten motorway concessionaires at the start of the decade, four were privately owned, their major shareholders being construction companies and banks, while the remaining six were owned by SEMs, often with very large, and sometimes total, public participation (state, regional and local). This position was modified in 1983, when three of the four private companies were taken over by the state-owned bank, the Caisse des Dépôts et Consignations (CDC), with the objective of being turned into SEMs (Holzmann and van der Bellen, 1986). Thus, by the mid-1980s, every motorway concessionaire except one was owned wholly or in part by public bodies.

The history of motorway concession in France is, therefore, one of substantial interpenetration of the public and private sectors, such that the very notions of conceder and concessionaire have become difficult to disentangle. At the start of the 1980s, when officials from the British and French governments met to discuss arrangements for building and operating a fixed link, the situation was becoming, if anything, more complex. This was due partly to the impetus given to public-sector involvement in all aspects of the economy by the election, for the first time in the history of the Fifth Republic, of a Socialist president (with a parliamentary majority), and partly to the natural evolution of a system which had been expanding for many years.

The Channel Fixed Link Concession Agreement

French negotiators of the Channel Fixed Link Concession Agreement thus brought to discussions with their British counterparts substantial experience of conceded infrastructure projects: since the mid-1950s, a number of French motorways have been built and operated by concession. From the other side of the Channel, British negotiators brought a determination that French experience should not be allowed to feed into the fixed-link concession: that almost all French motorways were in receipt of public assistance by the early 1980s was considered a poor precedent by the British. The resultant agreement reflects British commitment to a pure concession, in which financial risk is borne wholly by the concessionaire, although in practice it is given distinctive British and French interpretations.

The agreement which was signed by the British and French states and their joint concessionary partners, CTG and FM, is, then, a British purification of French practice. Throughout early discussions, French officials, with a developed public service view of transport, sought to write both government guarantees (in case of financial difficulty) and state regulation of fixed-link operations (particularly with regard to price) into the concession agreement. Clearly, these were twin aspects of a single policy, and each was rejected by British negotiators, who recognised that compromise on any aspect of the concessionary procedure would quickly undermine the private character of the project.

They were prepared to write political guarantees into a fixed-link concession, such that the situation of 1975, when the French government had been obliged to pay half the cost of British cancellation, would not be repeated, but they were not prepared to jeopardise the scheme's private character by countenancing state interference with, or subsidy of, the eventual concessionaire's commercial operations.

In every respect, then, the Channel fixed-link concession is designed to suppress French idiosyncracies, and to constitute a genuinely contractual relationship. The most fundamental revision of French practice is that the two states will bear none of the risks of fixed-link construction and operation: Article 1 of the fixed-link treaty states that 'The Channel fixed link shall be financed without recourse to government funds or to guarantees of a financial or commercial nature'. In return, the traditional French understanding that the conceder has the right to modify the terms of concession has been rejected, and regulation of prices by the two states has been denied, except in a situation of near- or actual monopoly. By extension of the logic of Article 1, continuity of public service provision is an obligation of the fixed-link concessionaire, but not of the two states. Should the concessionaire fail to meet its obligation, it does not transfer to the two states; and at the end of the 55-year concession, though the concessionaire is obliged to hand over the fixed link in full working order, the British and French states have no obligation to operate it.

The intentions of British negotiators in generating a genuinely contractual relationship were that fixed-link construction and operation be a properly private affair: no public assistance or guarantees should distinguish the fixed link from other commercial operations, and notably in this case, from Channel ports and cross-Channel ferry operators. The treaty and concession agreement establish an Intergovernmental Commission (IGC) to supervise fixed-link security, safety and environmental impact, and to assume responsibility for it in exceptional circumstances. Although, however, the IGC's control of these domains is complete, it was not envisaged, and has not turned out, that control be exercised in unnecessary interference with the concessionaire's commercial freedom.

In contrast to the messy character of French concessions, in which the conspiratorial nature of the nation's public life is extended to selected parts of the private sector, the concession agreement is thus pure in conception and application. In areas in which state co-operation is unavoidable, such as legislative enactment or compulsory land purchase, it is given. Outside them it is explicitly denied. Having provided the necessary legislation for fixed-link construction and operation, the stated intention of the two governments is that the fixed link operate as a strictly private undertaking.

Fixed-link concession and commercial franchise

There remains, however, a substantial gulf between the transport infrastructure concession, applied to the Channel fixed link, and the commercial franchise on the business format franchise (BFF) model, developed in the post-war period by companies such as McDonald's and Kentucky Fried Chicken in the United States, and by Body Shop and Tie Rack in the UK.

In each case — whether transport infrastructure concession or commercial franchise — an independent company is licensed to sell to the public a specific product or service. In return for the award of (local) monopoly rights, the licensee assumes part (or all) of the risk of goods or service provision. Thus, both systems conform to a definition of concession or franchising as licensed service delivery. When unpacked, this definition is found to incorporate the twin ideas which are central to the concept: that of a competent licensing authority in control of an exclusive product or service, and that of a financially responsible (or risk-bearing) licensee to whom goods or service delivery is delegated (sometimes for a fee).

However, despite their shared correspondence to this definition, the two types are in many senses dissimilar. Contemporary BFF systems are built on the concept of *system* leasing, whereby a patterned and branded product is licensed for sale in many local markets throughout the country, continent or world. The franchisor creates a marketing package over which it exercises continuing control throughout the period of the franchise.

By contrast, a transport infrastructure concession is non-patterned, non-branded, and virtually or actually unique. The licensor is able to retain a degree of control throughout the period of the concession, but such control is exercised not systemically or formulaically, but on the basis of a particular set of accords written into the concessionary contract.

In consequence, different forms of relationship, and distinct patterns of responsibility and control, are observed in the two types: relations are significantly altered by the amount of patterning inherent in the system. Although there is no linguistic basis for the distinction, for convenience the term 'franchise' will be reserved for BFF systems, and the term 'concession' for non-patterned transport infrastructure projects such as the Channel fixed link.[2] The key difference between the two is that under a highly formulaic franchise, the franchisor has extensive control over product specification and level of service delivery. Under an unbranded concession, the concessionaire exercises a significant part of this control.

This distinction should not be exaggerated. On the one hand, it is clear that franchisors' contractual control of product specification and service delivery cannot in practice always be total, particularly when a product is franchised extensively. Even a well-policed franchise is likely to permit some local autonomy and initiative. Nevertheless, franchisor product control is the basis on which commercial franchises are established, and deviation from this ideal, while possible at the margin, is unlikely ever to be substantial. McDonald's, which is franchised in many parts of the world but not (yet) in the UK, is patterned on a global scale, from the United States to the Soviet Union, and presents little evidence of franchisee initiative.

On the other hand, it is evident that the main parameters of the fixed-link concession were set by the British and French governments as the condition on which both the 1985 competition would be held, and the subsequent concession would be operated. Control of product specification and service delivery was at this stage retained by the joint conceders. However, at both stages in this process, it was only general guidelines which were issued by the two governments. Indeed, at the first stage they were sufficiently general to provoke a wide variety of tendered responses, ranging from road-rail or rail-only bored or immersed tunnels, to road-rail combined bridge-tunnel or bridge only schemes. The details were thus to be filled in, first sketchily by a series of tendering companies or consortia in the summer and autumn of 1985, and then, with great precision, by CTG-FM after January 1986. How these various processes were undertaken in Britain and France will be investigated here in terms of decision, and in terms of authorisation.

2.4 Divergent decision-making procedures in Britain and France

The fixed-link concession is both a British purification of French concessionary practice, and a different arrangement from commercial franchising on the BFF model. It thus stands in contrast both to the messy character of French concessions, and to the distinct character of the commercial franchise which, in pure form, is patterned throughout the world. The fixed-link concession is established on as pure a basis as possible, but its uniqueness distinguishes its purity from that of the BFF system. Within the single — and pure — framework provided by the Channel Fixed Link Concession Agreement, concessionary practice in Britain and France has been distinct.

When the fixed-link project was relaunched at the start of the 1980s, both governments and their officials sought advice from the private sector. In this practice there was nothing unusual: the 1960s and 1970s schemes had been developed in a similar way, and in other areas of the economy both governments rely to some extent on private-sector expertise.

At a very early stage, the fixed link looked like being a purely British project, as the new Transport Minister, Norman Fowler, invited proposals from construction companies and financiers without consulting his French counterpart. However, subsequent reports — among which that presented by a team of banks in May 1984 on the feasibility of private finance was particularly important — were commissioned by an Anglo-French team of officials. The members of this Franco-British Channel Link Financing Group, established in August 1982, were Midland Bank, and National Westminster Bank, Banque Indosuez,

Banque Nationale de Paris and Crédit Lyonnais, each of which was later to sponsor CTG-FM's proposals, and to emerge as a member of the lead group within the consortium of 200 lending banks which support the Eurotunnel scheme. From this time on, private-sector advice was sought in tandem — or at least in parallel — by the two governments. Where they differed markedly was in the amount of advice sought, and consultation undertaken, at the regional and local level. In France, this was extensive. In Britain, it was negligible.

Inter-governmental consultation in French decision-making

Consultation at the regional and local level in France started as soon as the project became a serious topic of debate at the beginning of the 1980s. The decision to reopen Anglo-French discussion of a possible fixed link was taken by Mrs Thatcher and President Mitterrand at their first summit, in September 1981, though (as was said in Chapter 1) it seems that real commitment on the British side was not generated for another three years. Certainly, there was little contact between civil servants and local government officers in Kent before 1985. In France, by contrast, working groups immediately went into action at both the national and the regional level. Moreover, at each level, consultation was undertaken with subsidiary levels, and with non-governmental bodies.

A key link between centre and periphery in France was the Prime Minister himself, Pierre Mauroy, who was at the same time Mayor of Lille and without doubt the leading Nord-Pas de Calais politician. Mauroy sponsored the Braibant Commission, which held both public and private meetings in important regional centres at the end of 1981, in an attempt to ascertain regional responses to a fixed link. This Commission reported directly to the Ministry of Transport. Simultaneously, the Nord-Pas de Calais Regional Council established a working group under one of its vice presidents, Daniel Percheron, both to investigate the likely impact of a fixed link on the regional economy and to outline measures necessary to its successful reception in the region. A report from the region's Economic and Social Committee, containing annexed contributions from chambers of commerce and agriculture, professional bodies and trade unions, fed directly into the deliberations of this group.

The resulting Percheron Report was adopted by the Nord-Pas de Calais Regional Council on 9 February 1982. A resolution passed on that day made a number of claims on behalf of the region: that there be a *rupture de charge* (or break in journeys, both passenger and freight) at the French end of a fixed link (to stimulate economic activity); that a public authority, with regional participation, be created to co-ordinate all modes of cross-Channel traffic (ferry, hovercraft, fixed link, air) and to determine price levels; that, again with regional participation, the fixed link itself be managed through the *procédure 'grands chantiers'* used by Electricité de France (EDF) in the construction of nuclear power stations; that a regional development plan be drawn up to take account of the impact of a fixed link on the regional economy; and that attention continue to be focused on the many regional investments which would be needed with or without a fixed link.

Some of these claims were later to prove unrealistic, when the French government acceded to the British demand that fixed-link construction and operation be conceded to the private sector. In these circumstances, the Regional Council could not, for example, directly participate in its management. However, the Regional Council continued to seek to influence the choice of fixed link, and, perhaps more realistically, to ensure that the needs of the Nord-Pas de Calais region be taken into account by French officials in their negotiations with the British. To this end, it commissioned Bechtel France to prepare a detailed report on the implications of a fixed link for Nord-Pas de Calais. When delivered at the end of 1985, Bechtel's *Impacts et perspectives pour la région Nord-Pas-de-Calais du lien fixe transmanche* (Conseil Régional Nord-Pas de Calais, 1985) gave substance to a set of regional demands, submitted to President Mitterrand on the day of the two governments' selection of CTG-FM at Lille town hall (Conseil Régional Nord-Pas de Calais, 1986b) and used as the basis for regional negotiations with the major tendering consortia.

In these negotiations, which took place mainly at the very end of the competitive period which closed with the selection of CTG-FM on 20 January 1986, the Regional Council was chiefly concerned to wring from each of the serious tendering consortia various commitments to the development of the region, its bargaining position being heavily dependent on its putative influence over the two governments' decision. By far the most important agreement, signed with all four main tendering consortia on 4 January 1986, was a commitment to target 75 per cent local (Nord-Pas de Calais) employment in fixed-link construction.

Private-sector initiative in British decision-making

In Britain, a very different decision-making procedure was enacted. In contrast to French practice prior to the January 1986 decision, where the Nord-Pas de Calais Regional Council was a major actor, and in some senses took the initiative in discussion of the local impact of a fixed link, in Britain in 1985 this initiative was taken mainly by the private sector, in the form of the major tendering consortia. In part, the private sector was propelled into this position by the British government, which made it a condition of fixed-link tenders that consortia engage in consultation at the local level, and demonstrate the local acceptability of their schemes. In part, the private sector was stepping into a void left by Kent County Council (KCC) and the implicated Kent district councils, none of which was prepared to adopt a positive negotiating stance with respect to the fixed link. This was either because — like Dover and Shepway — they were implacably opposed to the project, or because — like KCC — internal division prevented them from adopting a position on it.

Thus, in contrast to the situation in France, the pre-selection period in Britain was marked by more than three years (1981–4) of virtual inactivity outside the realm of central government and its private-sector advisers, and by less than one year (from April 1985) of local consultation sponsored by the tendering consortia. This consultation was only one of a number of activities undertaken in Britain chiefly by the private sector, and in France chiefly by the public sector. The other major activity for which responsibility was divided in this way was preparation of an Environmental Impact Assessment (EIA), which was governed in both countries by a recent EC directive on EIAs, but conducted very differently in each.

In France, although a measure of independent research was undertaken by each of the tendering consortia in 1985, the main fixed-link EIA was prepared during the public-inquiry procedure which followed selection of CTG-FM in January 1986. In Britain, more detailed EIAs were required to be included with tenders. This can be demonstrated by CTG-FM's submission, which was better documented than any other, but which nevertheless shows the vast demands placed on a serious fixed-link contender by the British government. Running to 18 volumes, CTG's EIA is a comprehensive source, and has turned out to embody a high degree of accuracy. In addition, CTG submitted documents to government on the tunnel's employment and other economic implications, and summarising environmental effects in the UK. The various specialist studies were undertaken by a series of individual consultants for CTG in 1985, and were brought together by a CTG technical co-ordinator. As the second tier of the study, Environmental Resources Ltd was appointed by CTG as environmental assessor to audit the work, to consider inter-relationships between specialist studies, and to provide an overview of significant impacts. A number of these studies were subsequently updated and published as Eurotunnel 'Baseline Studies' on environmental effects in the UK.

As part of the official evaluation procedure, the Department of Transport itself appointed consultants in August 1985 to make an independent appraisal of alternative fixed-link proposals (Land Use Consultants, 1986). This work drew on studies already undertaken by the four major promoters, and was essentially an assessment of work already done, rather than a full-blown independent study. In this way, the British government factored a large part of the cost of EIAs on to the private sector. It can also be argued that it thereby gained a better picture of the environmental implications of a fixed link than would have been possible had the project remained in the public sector. Each of these arguments — about cost (and therefore risk) and efficiency — can be applied more widely.

Consultation at the local level in the UK did not start with publication of the *Invitation to Promoters* in April 1985. Already, in February 1985, a Working Group of officials from each local authority directly affected by the Tunnel project — KCC, Ashford Borough Council, Dover District Council, and Shepway District Council — had been established to meet

CTG representatives; BR sent observers. In addition, *ad hoc* meetings on particular topics, such as employment and the environment, were held as necessary. However, after April 1985 consultation in Kent increased dramatically. CTG held an extensive series of meetings, both public and private, in the county, and supplemented its substantial local programme by wider discussions, with government officials, and with regional representatives in London and beyond. This process of consultation was consolidated in 1986 after selection of CTG-FM.

French and British decision-making compared: public versus private initiative?

The decision-making phase of the present fixed-link concession essentially came to an end on 20 January 1986, when the British and French governments agreed that a fixed link would be built, and that it would be done by CTG-FM. It is true that this decision had no legal or actual standing until after the successful parliamentary passage of legislation in the two countries, which ratified the draft fixed-link treaty, and confirmed the two governments' decision. However, difficult though this parliamentary passage sometimes was in the UK (it was not at all difficult in France), it did not, and could not, influence the *decision* taken by the two governments, except by rejecting it outright. This was not done. Thus, fixed-link debate turned to authorisation procedures.

Comparative assessment of influence and effectiveness in Britain and France in the decision phase is not easily managed. In large part, this is because this phase is not limited to the four-year period between the end of 1981 (Thatcher–Mitterrand summit) and the beginning of 1986 (choice of CTG-FM), but stretches back at least as far as the late 1950s, when serious work on the 1960s (which became the 1970s) scheme began. Technical and design studies undertaken at this time show a marked similarity to equivalent studies undertaken in the 1980s, and demonstrate the evolutionary nature of debate throughout this period: there was no sharp break at any point in it.

At each stage, private-sector promoters were important, both in generating design specifications at the start of the 1960s which have been largely replicated ever since, and in bringing to the attention of government at the start of the 1980s updated demonstrations of the feasibility of a fixed link. An indication of the continuity of private-sector contact with government is the submission to government in September 1981 of an outline project from Channel Tunnel Developments (1981) Ltd, proposing a 6 metre diameter single tunnel scheme (essentially the BR–SNCF mousehole). The two construction companies behind this scheme, Tarmac and Wimpey, were both to emerge later as sponsors of CTG, and the man who signed the 1981 submission, Tony Gueterbock, is still involved in the project in a major way, as public affairs manager at Eurotunnel. It is reasonable to say that no fixed-link scheme would have been chosen by the British and French governments in January 1986 had it not been for the sustained commitment of a small number of private-sector companies and consortia.

Within the context of this sustained commitment, the most important group of actors would then seem to have been the Anglo-French Study Group of senior officials, which to a large extent controlled debate in the 1980s, determining who should be invited to contribute to it, and on what terms. It was this group which commissioned financial and technical feasibility studies — from, for example, the Franco-British Channel Link Financing Group — and which judged the 1985 competition. It is this group which decided that the scheme which had been selected (by senior officials) in the 1960s remained more robust and viable than all others in the 1980s.

In these circumstances, it is hard to claim that the divergent consultation procedures adopted in Britain and France had much impact on the decision to build, and on the choice of CTG-FM. Despite its proactive stance, and its early and easy access to senior French officials, the influence of the Nord-Pas de Calais Regional Council on the January 1986 decision would seem to have been negligible. The key regional requirement of a fixed link, set out in the Percheron Report in February 1982, that there be a *rupture de charge* at the French end of the Channel Tunnel to stimulate economic development on the coast west of Calais, is crucially compromised by the through-train capacity of the Eurotunnel scheme. Equally, it seems likely that opposition in Kent, particularly in the port towns of Dover and Folkestone, would have been no more effective had it been voiced at an early stage through French-style consultation in East Kent. Indeed, the case against a fixed link was put to government during the decision-making phase, largely through the offices of Dover Harbour Board, and subsequently through Flexilink. It was, however, unable to counter a firm conviction at senior government level (from the end of 1984) that the fixed link was viable, desirable, and should be built.

Although in the decision phase of the present fixed-link project, initiative in Britain and France was distributed differently between the public and private sectors, this would not seem to have had an important impact on the January 1986 decision. Where divergent experience in Britain and France was very important was in subsequent phases of the project, where French institutional complicity and British institutional separation were to generate different policy outcomes.

2.5 Divergent authorisation procedures in Britain and France

By the time the present Channel Tunnel project reached the authorisation phase, it was in some senses two national projects rather than one international project. Highly divergent authorisation procedures were adopted in the two countries.

The French case: a streamlined public inquiry system

In investigating French authorisation procedures, it has to be remembered that the Tunnel raised little controversy on the French side of the Channel. In port towns, particularly Calais and Boulogne, it was viewed with hostility, and in small communes around the Tunnel portal it was awaited with apprehension. However, in the main, opinion was either indifferent to a Channel Tunnel, or in favour of it, seeing a fixed link as a chance for the regional economy, in structural economic decline for many years, to reposition itself in a rapidly changing Europe.

Clearly, this context is an important factor in a comparative analysis of authorisation procedures in Britain and France: in Britain, in contrast to the situation in France, the fixed link was widely considered undesirable by local residents. However, it is equally important to note two other points: that context is not simply 'given' but is created in part by the actions of public authorities; and that institutional practices are largely unaffected by context, and would have operated in much the same ways in Britain and France had the two situations been reversed.

In France, standard public inquiry machinery was activated once CTG-FM had been selected. However, for a project of the magnitude of the Channel Tunnel, this machinery itself contains a special procedure, the Déclaration d'Utilité Publique (DUP), which is used to subordinate private to public interests in the case of a project of strategic importance. In this way, economic or environmental concerns at the local level are in a sense trumped by the *utilité publique* of wider economic or environmental benefits.

It being clear that the Channel Tunnel might well fulfil the necessary requirements, the public inquiry which was conducted under French law was requested, as one of its main tasks, to determine whether the DUP should be promulgated in this particular case. After a brief, formal investigation of the project, which raised little controversy and was concluded in a matter of weeks, *utilité publique* was ascertained, and the Tunnel DUP was decreed on 6 May 1987. At this point, necessary compulsory purchase powers came into force, and the project was able to proceed.

The important point to be made about the French DUP is that it was in many respects simply the formal conclusion to a negotiated process of mutual economic benefit in which political and administrative elites had for a long time been engaged. The precise distribution of economic benefit will be investigated in later chapters.[3] The argument to be made here concerns that nature of French authorisation procedures, which were highly elitist, and very private. To see mayors from the rural communes directly affected by Channel Tunnel construction lined up to greet the arrival at inquiry of Nord-Pas de Calais Regional Council president, Noël Josèphe, was to witness the nature of relations between the regional centre and its periphery (which in many ways replicate those between Paris and the regions). Instructions had evidently been issued, and were being followed. However, what was equally clear was that lengthy and hard bargaining had gone on behind the scenes to produce this façade of unity.

Compulsory purchase powers were also exercised without difficulty in France, partly no doubt because French compensation procedures are both more flexible — permitting consolidation and redistribution of non-acquired but disturbed holdings — and more generous — allowing compensation at over 100 per cent — than their British counterparts. However, again elite accommodation is the main explanation for the smooth operation of French planning mechanisms. The issue of compulsory land purchase, and the ease of its resolution in France, will be investigated in greater detail in Chapter 6.[4]

Unusually in the case of the Channel Tunnel, a further special procedure was activated. This, the *procédure 'grands chantiers'* (PGC), had never previously been used except in cases of public works projects, and then only in the specific case of nuclear power station construction by EDF. Developed by EDF in the 1970s, and refined by it in the 1980s, the PGC was adapted to the requirements of the Channel Tunnel. It was put not under the control of the Nord-Pas de Calais Regional Council, as requested in the Percheron Report, but under the control of central government, through the prefectoral system. The *grand chantier* co-ordinator is appointed by the national regional planning agency, DATAR, and placed under the authority of the prefect of the Pas de Calais (Guillot, 1988).

In some respects, the PGC is the complement of the DUP, though an important distinction between a legal process (DUP) and a merely managerial technique (PGC) needs to be borne in mind in making this statement. Where the DUP subordinates private to public interests, the PGC attempts to ensure that local *implantation* of, in this case, the fixed-link project is managed as sensitively and as effectively as possible. The co-ordinator cannot direct the commercial activities of either Eurotunnel or TML, but he can and does act in collaboration with them in a number of areas. In particular, he brings together the Channel Tunnel promoters and local interests — represented by both elected persons and non-elected bodies — to consider jointly employment and training requirements and environmental matters.

As a consequence of the smooth operation of the public inquiry, the legislative machinery used in France to plan the fixed-link project was minimal. Debate in the National Assembly included several contributions from regional *députés* who felt the need to record the many accompanying measures required by Nord-Pas de Calais if it was to profit from the fixed link, but it was not marked by dissent. The fixed-link treaty was ratified on 23 April 1987 on a unanimous vote in the National Assembly, and on 3 June in the Senate.

The British case: a hybrid bill procedure

In Britain, the 1980s Tunnel project was controversial, though in some respects less so than its accompanying high-speed rail link (from the Tunnel mouth to central London) was to be only some four or five years later. A similar pattern of dissent had been predicted in the 1970s. Furthermore, it was controversial not only in itself but also in the means used to authorise it. This was a hybrid bill procedure, which took the Parliamentary route, rather than a public-inquiry procedure, which would have found a passage through local hearings to a final arbiter, the Secretary of State for the Environment.

About this procedure, it is important to state at the outset that, just as the public inquiry and DUP were normal French practice, so the Parliamentary device of a hybrid bill made no departure from normal UK practice. The 1970s scheme had also used a hybrid bill, with the government as promoter. The sole significant difference in the 1980s was that, having secured powers to develop a fixed link, the government then made them over to a private-sector concessionaire.

It might further be noted that a similar procedure is customarily used to authorise the building of railways in the UK: the classic nineteenth-century device is a private bill. To this day, it remains the instrument used by BR for all railway construction, and a number of such bills are deposited in Parliament each year. The difference between a private bill and the hybrid bill used to authorise construction of the Channel Tunnel is that a private bill is of strictly individual application — it contains provisions like the acquisition of a person's land — whereas a hybrid bill has both private and public elements. The public parts of the Channel Tunnel Bill were chiefly its international aspects — the treaty with France which required ratification — its setting up of an intergovernmental body, the IGC, and its prohibition of public subsidy of the project.

Thus, the hybrid bill for Channel Tunnel construction was completely within the normal operations of British Parliamentary procedure. This is not to say that it was, therefore, either uncontroversial or appropriate. The standard criticism made of it — forcefully in Kent — is that it denied Kent residents a number of rights which they would have had under the alternative mechanism of a standard planning application under the Town and Country Planning Act 1971, and an ensuing public inquiry (which would have been unavoidable given the size of the Channel Tunnel project).[5] In particular, it denied petitioners to the Parliamentary Select Committees set up to consider the bill's provisions the right to object to the Channel Tunnel in principle: the principle had already been decided by the second reading of the Channel Tunnel Bill in the House of Commons on 5 June 1986 (309 votes in favour, 44 against: most Labour MPs abstained).[6]

This is an important criticism, but it tends to exaggerate the merits of the British public-inquiry system, which is itself now largely discredited in many people's eyes. It is true that a public inquiry may allow objection in principle. There is, however, no guarantee that such objection will be heeded, either by the presiding inspector, or by the Secretary of State for the Environment, who takes the final decision in all public inquiries, and who does not have to take account either of any views expressed at inquiry, or of the views of his own inspector.

Nevertheless, the central criticism which can be made of the hybrid bill procedure used in Britain to authorise Channel Tunnel construction remains that it was not sufficiently democratic. This objection requires careful examination. To begin with, Select Committees of both Houses of Parliament invited petitions from all local — but not, because of the *locus standi* condition, general — interests, and in open hearings heard all petitioners who wished to be heard. The Commons Select Committee received 4845 petitions, the Lords 1457; because of these vast numbers, petitions were heard in groups (according to similarity). The Commons Select Committee sat for a total of 36 days in public hearings, and listened to over 220 hours of oral evidence. Furthermore, it did not conduct its hearings solely in Westminster, but, for the first time ever, left London to travel to Kent for sessions in Hythe (16–19 September 1986) and Dover (22–23 September 1986). In addition, petitions were made directly to decision-makers (rather than to an inspector), and, if considered significant, could thus be incorporated in a negotiated resolution of difficult problems. By contrast, a public inquiry is very bad at securing negotiated compromises over points of detail: a protester simply presents an objection, which is noted by the inspector (and may or may not be heeded by the Secretary of State).

This point about negotiation is, however, most relevant not to individuals, who have little negotiating strength, but to the key public and semi-public institutions, such as in particular the local authorities, which secured a number of modifications of Eurotunnel's original proposals through the hybrid bill procedure. Perhaps, then, this procedure is best described, not as a victory for democracy, but as a victory for corporatism. This notion contains the twin ideas of elite negotiation and mass exclusion which have both been witnessed in some degree by British experience of the Channel Tunnel.[7]

At this point, British practice seems to be in line with French. Apparent similarities between the two are, however, highly misleading. Crucially, very different forms of elite negotiation were conducted in each country. In Britain, the focus of debate was planning consent and control, and the mechanism of elite negotiation was a lengthy and often very public hybrid bill procedure. In France, by contrast, the focus of debate was economic benefit, and the mechanism of elite negotiation was an essentially very private public inquiry procedure.

Comparative assessment of authorisation procedures in Britain and France

The difference between British and French authorisation procedures that is revealed by the Channel Tunnel experience requires careful explanation. The difference is, of course, not simply in the nature of the procedures, but also in the nature of the problems with which each was required to deal. If anything, these problems were more divergent in the case of rail links to the Tunnel in Britain and France, than in the case of the Tunnel itself, and this subsequent experience may be used to reinforce lessons learnt from planning the Tunnel.

The ultimate difference between authorisation procedures in Britain and France would seem to be that one, the French, is permissive (to the developer), whereas the other, the British, is obstructive. French plans to build TGV Nord between Paris and the Tunnel mouth met some resistance in the Lille *métropole*, where the TGV enters the city and runs to a station in its centre, but these did not prevent the smooth passage of planning applications, and a DUP on 29 September 1989. In part, this was because resistance was extremely localised and limited, being confined largely to the Lille suburbs of Lambersart and Saint André. In part, it was because French law permits compensation at over 100 per cent of market value, and thus allows opposition to be bought off through financial packages which may actually be to the advantage of those expropriated. In part, it was because political will was sufficient to overcome opposition which was felt to be against the general interest.

By contrast, BR's plans to build a Channel Tunnel rail link from Cheriton to central London met extensive resistance along the entire route. As a result, BR was forced, first, to incorporate extensive environmental modifications — including a 29 km tunnel under south London — in its plans (thereby raising the estimated cost from £1.2 billion to £3.5 billion); second, to abandon these plans because they could not be profitably funded; third, to seek public subsidy of the route because even a modified version would not make a market return; fourth, to abandon all plans because of government refusal to meet the cost of environmental protection on the route; and fifth, to devise new plans through consultation with other potential developers. Clearly, this catalogue of problems cannot be ascribed merely to British authorisation procedures, which have never formally been activated by rail-link plans.

Nevertheless, these procedures are partly implicated in the British rail-link fiasco. This is because the shadow of their provisions has fallen over the entire rail-link debate. Important in this respect have been the compensation arrangements which they embody. In the case of expropriation by public bodies — including nationalised industries, such as BR — these arrangements are governed by the Land Compensation Act 1973, which provides for full and fair compensation (up to 100 per cent), but prohibits compensation above 100 per cent.[8] It thus offers little incentive to objectors to acquiesce in their own expropriation, a fact which was well understood by rail-link protestors in Kent.[9]

More important, however, than the technicalities of British compensation arrangements has been the lack of political will behind the project. Having failed the market test — in large part because of politically inspired environmental safeguards — the project had no reservoir of political commitment to fall back on, and was allowed to drift into extinction. A similar fate befell the 1970s project.

When the full circumstances surrounding both the Channel Tunnel and rail links to it are considered, authorisation procedures in Britain and France thus account for only a part of observed differences. That part, has, however, been very important. Moreover, it reflects fundamental differences in operation between the British and French political systems. At each point in the development of the 1980s Channel Tunnel project, these differences are revealed.

In the early stages of the project's development, when it was above all an affair of state, the work of two national administrations is visible. It is true that French negotiators wished to widen the circle of negotiation, to encompass officials from the Nord-Pas de Calais region, but this proposal was blocked by British officials, who had no intention of increasing the complexity of discussions by involving officials from Kent. French concern to maintain *parité* prevented a break with the British line, though it did not diminish the amount of central–local negotiation which was conducted outside, and to some extent alongside, the series of binational talks. The Braibant Commission, sponsored by both Socialist Prime Minister (and Mayor of Lille), Pierre Mauroy, and Communist Transport Minister, Charles Fiterman, conducted extensive discussions in Nord-Pas de Calais, and was well appraised of the region's hopes and fears with regard to the fixed link. The paltry contact established between central and local government officials in the UK in no sense compares with the developed French links.

In part, a basic congruence between local and national perspectives on the Tunnel in France facilitated contact and negotiation between different levels of government. However, a more fundamental point may be made about this difference in institutional experience. This is that the French system is structurally more integrated than the British. The *cumul des mandats* is undoubtedly an important aspect of this integration, and in the case of the Channel Tunnel is seen most vividly in the person of Pierre Mauroy: Prime Minister of France 1981–4; president of the Nord-Pas de Calais Regional Council 1974–81; Mayor of Lille since 1971. The case of Mauroy, *cumulard sans pareil*, is spectacular. It is at once both decidedly atypical, and fully revealing of the French system of government, which operates in the knowledge that the centre cannot take local initiatives without first lining up local *notables*. This would have been the case had the Tunnel been desirable or undesirable in Nord-Pas de Calais. Indeed, it is difficult to say whether construction of a fixed link would have been possible if the relevant French localities had not been at least prepared to tolerate the project.

In Britain, no such worries were expressed in the initial stages of the fixed link scheme: British officials were almost wholly independent (if not ignorant) of local wishes. Worries have, however, surfaced subsequently, as Kent residents have found themselves at the receiving end of a government scheme which in many respects and in many parts of the county is considered undesirable. The underlying discord between centre and periphery can account for some of this mutual isolation, but not for all of it. It is a more general feature of the British system that there is very little contact between the centre and localities at any level, whether political or administrative. Backbench Conservative MPs in Kent had little contact with government in the project-development phase of the fixed-link scheme; KCC was very slow to take a position on the issue, and at periodic meetings with government officials had little to contribute; its officers, in meetings with officials from the Department of Transport, were unclear what line their members would ultimately take. Aspects of the intergovernmental relations described here are clearly Tunnel-specific, but the point is more generally applicable to most government policies, unless the policy impinges directly on local government, in which case it tends to be discussed with the various local government associations (which are often divided).

The difference between the British and French systems is, then, that in Britain the government tends to produce proposals to which the localities react, whereas in France, policies are agreed with the localities before they are made public. In the case of the fixed link, in France national administrators engaged in a process of negotiation with relevant local authorities which paralleled negotiations with the British; in Britain, contact between centre and periphery was limited in essence to periodic exchanges of information between the Department of Transport and KCC.[10]

This disjuncture made development of a negotiating position with regard to the fixed link a great deal easier on the British side than on the French. In Britain, Department of Transport officials were able to resolve through a few telephone calls — or even through a single telephone call to the Secretary of State — matters which in France were the subject of protracted negotiations. However, it is not clear that in the long term British practice is more efficient. Certainly, in the case of the fixed link, difficult questions which were evaded in the development phase of the project have since emerged to disrupt later phases. The point must be emphasised that in many respects this was unavoidable. KCC's inability to formulate a position on the fixed link is understandable; and it is certain that an attempt to address all the Tunnel's implications at the outset would have wrecked the scheme. This is particularly true of a Channel Tunnel rail link, which, through rapidly escalating costs, had been abandoned in the 1970s, and which was (and remains) highly undesirable to a significant, and vocal, minority of Kent residents.

2.6 Conclusion

More general lessons about British and French government must, however, be drawn from the Tunnel experience. The clear strength of the British system is its degree of co-ordination at the centre of government. The initial British negotiating team, established at the turn of the year 1984/85 to issue (with its French counterpart) the *Invitation to Promoters*, was assembled by telephone in the course of a morning, and confirmed without debate by Nicholas Ridley, then Secretary of State for Transport. This team comprised officials from the Department of Transport, Treasury, Foreign Office, Cabinet Office and Downing Street

Policy Unit. The same process took several weeks in France, having, as it did, to negotiate the complex web of inter- and intra-departmental rivalries and jealousies, disputes between ministerial *cabinets*, and demands from the Nord-Pas de Calais region. The final French team was ratified by the Council of Ministers, and published in a minsterial decree. Thus, in the time it took the French to assemble a negotiating team, the British were able to do a large amount of preparatory work, and the *Invitation* was issued, as planned, at the start of April.

Further French delay hampered subsequent assembly of evaluation teams. On the British side, it was assumed that the existing groups would simply progress to the next stage, and officials from the Departments of Environment and Trade and Industry were co-opted to strengthen the team. By contrast, on the French side, the existing team's remit was held to have expired with publication of the *Invitation*, and it was not until the start of October 1985 that the setting-up of a French evaluation group was announced. There was further delay before its composition was announced. Bids were due on 31 October.

Again, then, in the time it took the French to assemble for the start of the next stage in the process, British officials were able to get important work done. In this period, this work comprised detailed planning of the terms on which competing bids would be appraised, engagement of technical consultants and financial advisers, and preparatory work with these consultants and advisers. As a result, French delay was again not fatal, but it is no exaggeration to say that only the speed and flexibility of the British system kept the fixed link project to timetable in 1985. It is even possible that the entire project would have collapsed had it not been for the ability of British officials to make up for delay on the French side. In any judgment of the two systems on the basis of the Tunnel experience, the events of 1985 must be borne in mind.

That judgment must therefore be that the French system is excessively formal and legalistic, and in this sense inferior to the more collegiate system, held together by the Cabinet Office, which exists at the centre of British government. By this is meant that the British system is clearly quicker on its feet than the French, and, in consequence, may have saved the Channel fixed link from one more in a long line of collapses in 1985. This is no small achievement.

A number of reasons for French inflexibility may be cited, among them the fact that the French simply are in some senses more legalistic than the British in their approach to public administration. However, one of the key defects of the French system which is revealed by the Tunnel experience is more narrowly institutional. This is the role of ministerial *cabinets* in acting as a filter — even in some cases a barrier — between senior officials and ministers. In Britain, officials know their ministers' minds, and can act in full confidence of subsequent approval. If in doubt, they have immediate and full access. None of this is automatically the case in France, where officials frequently have to negotiate *cabinets* in order to reach ministers. Certainly this was the situation during discussion and evaluation of the present fixed-link scheme. Throughout 1985, and indeed right up to the selection of CTG-FM in January 1986, the leading British official, Andrew Lyall, was closer and had better access to Auroux, the French minister, than his French opposite number. This is a ridiculous state of affairs.

Thus, the British system has clear merits when it comes to directing a process such as that which culminated in the decision of 20 January 1986, its main attributes being the speed and flexibility which are equally clearly deficient in France. However, whether these are the ultimate virtues of a central administration must be questioned by experience of the Channel Tunnel — for the involvement of government officials in this project did not end with selection of CTG-FM.

Instead, many further issues which required negotiation with regional and local interests in particular remained to be settled, and were settled in divergent ways in Britain and France. Once these issues are also taken into account, the relative merits of the two systems appear in a different light, for French central government has links into the periphery which cannot be matched by British. Furthermore, these links are to a large extent established by the very ministerial *cabinets* which, as demonstrated by the Tunnel experience, hamper speed and flexibility in official action. In these circumstances, the strengths of the two systems must be reassessed. Difficult though it is for the British to admit, it may be that the intricate and highly politicised web spun by French government is ultimately more efficient at generating acceptable solutions to the complex problems of modern government than are the gentlemanly, rather proper, and decidedly distant workings of British government. It is not clear that being *au-dessus de la mêlée* is the right position for a government official.

The point is, that once (finally) set up, the French team of negotiators was able to check off decisions at all important levels of government as they were debated and taken. This was not done in the UK. Thus, when formal authorisation procedures were subsequently set in motion, many important French interests had already fed their needs into the proposal which was to be considered, and did not object to it. In Britain, the more correct and visible workings of the planning system prevented this. Though tortuously established, the French team thus co-opted a broad range of interests and was able to proceed without significant difficulty through the many phases of the Tunnel project which were to follow the initial government-dominated phase. The British team, assembled by a morning of phone calls and plugged directly into 10 Downing Street and the Cabinet Office, exercised real executive power, but co-opted few — if any — interests outside central government. Though crucial to project survival in 1985, the British system was thus to be less well equipped than the French in subsequent years.

Unsurprisingly, these institutional differences have had an important influence on the success of the Channel Tunnel scheme in Britain and France, in both narrow and broad terms. This entire debate will be investigated at greater length in Part Two. However, feeding into debate in each country has been a quasi-scientific debate of cost and revenue forecasts, which will first be considered.

Notes

1. See King (1975) for a representative expression of the 'overload' thesis, which noted that demands on the British government had increased at the very moment when its capacity to address them was constrained by the number of dependency relationships in which it was involved. Overload was said to generate ungovernability.
2. The *Oxford English Dictionary* defines concession as 'A grant by government of a right or privilege', and franchise as 'A privilege granted by the sovereign power to any person or body of persons'.
3. See in particular Chapter 6.
4. See pp. 107–8.
5. The Labour Party promised to institute such an inquiry if it won the 1987 general election.
6. This is the *locus standi* condition imposed on petitioners during the Parliamentary Select Committee hearings, which prevented any discussion of issues not directly linked to personal interests.
7. A criticism of BR's use of private bills should be noted here. This is that it only allows consideration of the scheme proposed by BR. Genuine consideration of alternative routes is ruled out because these would involve different private interests, and would thus require a fresh bill. This point applies equally to the hybrid bill procedure used to authorise the Eurotunnel project.
8. There is also provision for a disturbance allowance, currently of up to £1500, which may be paid on top of 100 per cent compensation. However, given the present market value of houses, particularly in the South of England, this cannot become a disguised means of compensating at over 100 per cent.
9. This is a question which has been addressed by the House of Commons Environment Committee. In its report, *Radioactive Waste* (House of Commons, 1986, Vol. 1), it noted the flexibility and pragmatism of French compensation procedures (Annex 3), and recommended that the Department of the Environment give renewed consideration to British compensation arrangements (para. 245).
10. This difference will be explored at greater length in Part Two.

3 *The political economy of the Channel Tunnel*

Private funding of the Channel Tunnel project has precluded authoritative analysis either of the internal profitability of the Tunnel scheme itself, or of the external costs and benefits likely to be associated with it. On occasion throughout the 1960s, 1970s and early 1980s, government officials have conducted studies of the likely return on various schemes, but, once the decision to concede the fixed-link project to private enterprise had been taken, this work was viewed as commercially sensitive, and has not since been released.

In the absence of authoritative work, competing studies of both the private and the social returns which can be expected from the Tunnel project have been undertaken by a number of bodies. Many of these are interested parties. In the early 1980s, competing private-sector consortia issued assessments of the internal profitability of competing fixed-link schemes, which were countered by the port and ferry industries, often operating through Flexilink. Since January 1986, this series of studies has been extended and refined by Eurotunnel, as part of its campaign to raise funds from banks and private individuals. In this period, its future competitors in the ferry industry have generally preferred to keep their calculations under wraps. The indications are that over time the ferry industry has become more optimistic about its chances in a market dominated by Eurotunnel.

Since 1986, wider studies of the regional economic impact of the Channel Tunnel have been undertaken, again in many cases by interested parties. In Britain in the late 1980s, production of these studies was at a peak, as many regions sought to determine the impact of the Tunnel on their economies. In almost all cases, the intention was not only to determine how regions could themselves best respond to the challenge of the Tunnel but also to put pressure on the government to fund transport links — particularly by rail — to regions distant from the Tunnel, and in many ways cut off from it by the major bottleneck of London and the over-crowded South-East. In the space of a few years since 1985–6, the regional consensus has shifted from denunciation of the Tunnel and associated works on the basis of perceived negative impacts, to enthusiastic promotion of both the Tunnel itself and infrastructure links in south-east England as a means of linking distant economies to the rest of Europe. A number of independent analyses of the regional economic impact of the Channel Tunnel were also undertaken in this period.

The purpose of this chapter is to present an objective analysis of the Tunnel's economic calculus, as a means of assessing the real — as opposed to the stated — return likely to be generated by it. The chapter will confine itself to the internal dynamics of the Tunnel scheme itself, and will leave consideration of its wider regional impact to Chapter 10, in which the broader consequences of the Channel Tunnel scheme are reviewed.

3.1 Debate of the economic calculus

At the centre of debate about the Tunnel's economic calculus has been the issue of forecasting the likely costs and revenues of the Eurotunnel scheme. Each will be influenced in important ways by factors which are entirely out of Eurotunnel's control: in the case of costs, by inflation and interest rates, particularly in the UK and France, but also in the world economy; in the case of revenues, by the general level of economic growth in the economies which form the Tunnel's extended hinterland. Each will also be influenced in important ways by factors which can only indirectly be controlled by Eurotunnel: in the case of costs, by the efficiency with which TML manages to build the Tunnel; in the case of revenues, by the resolution of the rail-link problem in the UK, and the completion of road links in France. Each list could be extended.

The uncertainty created by this set of circumstances provides a situation in which a great variety of potential scenarios is plausible. Indeed, it was partly for this reason that the British and French governments — and particularly, it is thought, the British — selected the CTG-FM project in January 1986. Being, in essence, the scheme which had been studied and chosen in the 1960s and 1970s, this project was thought to be financially more secure than any of its rivals, and in particular to be less prone to cost overrun from unknown factors. In the event, costs had already spiralled sufficiently at the end of 1989 (from an estimated £4.8 billion in 1987, to an estimated £7 to £8 billion only two years later) to provoke a major crisis at the heart of the Channel Tunnel project in a dispute between Eurotunnel and TML. It is by no means certain that this problem will not arise again.

In these circumstances, it is important to be clear about the economic bases of the project. In principle, the main consequence of making the Channel Tunnel project a purely private-sector venture — with no government guarantees, except of a political nature — is to shift risk from the public to the private sector. In practice, the real locus of risk has yet to be determined: only if the Tunnel is actually threatened with bankruptcy will a definitive location be possible. However, this section will operate on the assumption that risk has genuinely been conceded to the private sector.

In the case of the Channel Tunnel, risk arises on two sides: from unforeseen problems in construction which lead to unexpected increases in costs; and from the problems of forecasting revenues for a new mode of transport for an initial traffic up to seven years in advance of its operation, and for subsequent changes in traffic over a further 20–30 years. Traffic revenues depend both on overall levels of traffic between the UK and continental Europe, and on the Tunnel's share of that traffic. In turn, the latter will depend both on the Tunnel's own pricing strategy, itself dependent on costs, and on the degree of competition from alternative modes of transport, ferries and airlines.

This chapter seeks to establish a framework for understanding the way in which these various elements fit together, and to draw some a priori conclusions about the likelihood of a return on the project which is acceptable to the private sector. It investigates progress to date, and assesses the present chances of the project making an economic return. It should be stressed again that only the internal profitability of the Channel Tunnel is investigated in this chapter; any external or induced costs or benefits associated with the Tunnel are irrelevant, unless they can be internalised in revenues to Eurotunnel, the Channel Tunnel operator.[1] The following sections look in turn at cost structures, traffic forecasting, pricing, and competition, and present an overall investment appraisal.

3.2 Cost structures

There has always been a tension in British official thinking about the appropriateness of using the private sector in public-utility provision. On the one hand, the public sector, both in Britain and abroad, has shown itself to be a poor manager of large construction projects, with cost and time overruns typical. The Humber Bridge, planned to cost £19 million in the mid-1960s, eventually cost £120 million when opened late on 24 June 1981. The Thames Barrier, provisionally budgeted at £13–18 million when planned in the early 1960s, was subsequently forecast to cost £110.7 million when construction started in October 1973, and to take just over five years to build. When completed ten years later, the final cost was put at £440 million. It is estimated that some 70 per cent of the cost overrun was due to inflation, 15 per cent to poor productivity, 10 per cent to construction difficulties, and 5 per cent to design enhancements (Morris and Hough, 1986, p.76). In the USA, the Shoreham nuclear plant near New York, budgeted at around $250 million, was finally completed at a cost of $5.1 billion. It was subsequently closed by the state governor.

On the other hand, the public sector has access to cheaper funds than does the private, such that in theory the total financing cost of a public-sector scheme should be less. Historically, this view has been taken by the Treasury, and has prevailed, thereby restricting private-sector involvement in large construction projects (and restricting public-sector investment through tight controls on the public sector borrowing requirement).

It should be noted that this situation is changed somewhat when there is an equity stake in a project, since this distances the government from risk. However, it does highlight the importance of separating investigation of construction costs and total financing costs.

A second issue is the fact that most investment in projects of this magnitude is undertaken by organisations which have previous experience of managing large construction projects, which have a financial track record against which both investors and lenders can assess the degree of risk, and which, most importantly, are earning revenue during the construction period. In this respect, Eurotunnel is unique as a purpose-built single-activity joint-venture company which cannot earn revenue against its investment until that investment is complete in every respect. By contrast, toll roads, for example, can be built in stages, thereby spreading the risks of both construction and traffic forecasting. The Channel Tunnel is in the more awkward position of being a single, indivisible project, although there is a possibility of staging construction (Gwilliam, 1983).

In operation, the Channel Tunnel is, however, the classic infrastructure of the type originally discussed by Dupuit (1844), with a large sunk cost, zero opportunity cost of capital, and low (virtually zero) constant marginal cost of operation up to capacity. In theory, user charges for this type of investment should be set at marginal cost, the cost of initial investment not being borne by current users. Capital costs should only become relevant when demand reaches capacity; at this point, charges should be increased both to ration demand and to finance investment in further capacity. However, economic theory here presupposes public-sector ownership, and public-sector assumption of capital costs as a means of creating wider non-user benefits. In the case of Eurotunnel, where construction costs are borne entirely by the private sector, standard theory does not therefore apply directly.

The problem posed for a private-sector Channel Tunnel is thus that the need to repay debt incurred in Tunnel construction implies a user price significantly above the marginal cost of operation. This will limit the competitive flexibility of Eurotunnel, and will also reduce the potential social value of the Tunnel since a price above marginal cost transfers benefit from users to equity and debt holders, as well as creating a deadweight loss. In any full calculation of the desirability of conceding infrastructure projects to the private sector, allowance for this deadweight loss, and for the distributional considerations noted here, must be made.

Thus, the three basic parameters to the Channel Tunnel's cost structures are, first, the ability of Eurotunnel to manage costs; second, the ratio of equity to debt (and the consequent distribution of risk), and the total financing cost; third, the extent to which price exceeds short-run marginal cost.

As was noted above, large construction projects are notorious for cost and time overruns (Dufloux and Margulici, 1984; Morris and Hough, 1986). There are several reasons for this. First, for financial reasons, both public and private developers only undertake detailed design studies and costings after a decision to proceed has been taken. Partly this reflects the way in which decisions are timed: large projects are only considered to be viable within acceptable bounds of risk when they are actually needed. It is thus difficult to provide a convincing investment decision at the right stage in advance of need. Partly it reflects private-sector reluctance to stake risk capital on design in advance of securing a concession. In the case of the Tunnel, the result was that although many years — stretching back to the late 1950s — had been devoted to outline consideration of the scheme, only a very brief period was allowed for detailed design studies in advance of construction.

At the same time as it was developing a detailed design brief, Eurotunnel was also engaged in seeking parliamentary authorisation for its proposed scheme. In France, this was not a significant constraint on management time. In the UK, however, it made considerable demands on the company. The 'fast-track' approach to design which Eurotunnel adopted thus meant that within 7.5 years the project will have moved from outline consideration to operation. However, it also meant that a number of design problems were not identified at the start of the project, and that provision for them was, therefore, not made in Eurotunnel's initial costings. The real extent of this problem was at the heart of the dispute over costs between Eurotunnel and TML at the end of 1989, with TML arguing that cost increases were chiefly due to deficiencies in the initial design brief and costings, and with Eurotunnel taking the opposed line that they were chiefly due to TML inefficiencies in following a perfectly satisfactory brief. The independent *maître d'oeuvre* appointed to assess these competing claims found largely in favour of Eurotunnel, but the timing and extent of the design phase remains a problem in projects of this magnitude.

In the case of the Channel Tunnel, a further complication arose from public surveillance of the project through the Intergovernmental Commission (IGC), which has the right — and the duty — to consider operational aspects of the Tunnel scheme, such as safety, only after detailed design has been completed. Through a series of *avant projet* submissions, Eurotunnel and TML are given authorisation for designs which do not form part of the concession agreement signed with the two states in March 1986. The concession agreement itself contained merely a general outline of the scheme; *avant projet* submissions fill in its details. In theory, the IGC can reject any design on grounds of safety, security, environmental acceptability, and so on, though in practice constraints on its power must exist. Nevertheless, a further element of risk, and thus of potential cost and time overrun, is introduced to the Tunnel project by this regime. It should be noted that the critical decision over non-segregation of car and coach passengers was only given in December 1989 (Channel Tunnel Safety Authority, 1990), several months after an initial £600 million contract for rolling stock had been placed by Eurotunnel in July 1989[2].

A second reason for cost and time overruns on large construction projects is that very often they are developer-led. In these circumstances, it is extremely difficult to identify a clear client–contractor relationship at the heart of the project. Public-sector projects have been plagued by this problem, and redesign of them has frequently been difficult to control: the client, a non-specialist in the field, is viewed as having unlimited resources given its initial justification of, and commitment to, the project on the basis of need. The particular problem in the case of the present Channel Tunnel project — and of all previous fixed-link schemes — is that its original promoters were construction companies and banks which sought their main return from construction of the Tunnel, rather than from its operation. Although the embryonic Eurotunnel acted quickly to distance itself from this core group of founder shareholders, it was bound to be difficult for a new organisation to negotiate fair contracts with such an experienced group of contractors. With hindsight, a preferable arrangement might have been a series of separate contracts for separate sections of the work, rather than the single design, build and commission contract which the nascent Eurotunnel let to its founder shareholders in the new guise of TML. It also seems that the banks which financed Eurotunnel in its early stages, and which now form the core of the syndicate of 200 banks which support the scheme, acted too hastily, and without sufficient regard for the inherent problems of a contract of this kind.

A third source of cost escalation in the course of a construction contract is the pressure tendering companies are initially under to cut cost estimates to the bone, in order to make a successful tender. Concession of the fixed link to the private sector did not evade this problem, since competing fixed-link consortia knew they would be judged on financial viability, which encouraged them to minimise margins until the concession had been secured. Delays arising from the parliamentary process which authorised the two governments' selection, or difficulties in financing the Eurotunnel project, were then exploited as 'causes' of cost increase.

Fourth, an imbalance in the client–contractor relationship is generated by the fact that there comes a point in any large construction project when the cost of time overruns (in the form of lost revenues) is more damaging than direct cost increases. Thus, power in the client–contractor relationship remains skewed in favour of the contractor throughout the construction and commissioning period, with the result that cost escalation is always possible.

Each of these factors has helped to escalate the cost of the Eurotunnel project. Initial (1985) estimates put construction costs at £2.3 billion (in 1985 prices, as are all prices in this paragraph). The 1987 Eurotunnel prospectus for Equity III revised this figure — upwards — to £2.7 billion, which was said to imply a total financing cost of £4.8 billion. Since the 1987 prospectus, there have been two rounds of dispute with TML. In 1988–9, the major issue was the schedule for completion of the Tunnel: planned opening day was moved from 15 May to 15 June 1993. Estimated construction costs were increased to nearly £3 billion, and total financing costs to £5.4 billion. In 1989–90, a much bigger dispute arose over the cost of lump-sum works, such as terminals, rolling stock, and equipment. The key aspect of this set of cost increases was that the rise in total financing costs to a minimum of £7 billion took Eurotunnel over its total financing limit, secured by Equity III at £6 billion, and put it in technical breach of its credit agreement with the banking syndicate.

The issue came to a head when the banks' independent technical adviser suggested that even TML's latest figures were lower than the likely total financing cost of the project, and that their figure of £8 billion did not provide a basis for additional funding. The independent *maître d'oeuvre*, called in to arbitrate between Eurotunnel and TML, found largely in favour of Eurotunnel, and an interim agreement was signed in January 1990 which resolved some difficulties, and provided Eurotunnel with a firmer basis for increasing its total financing limit by £1.2 billion to £7.2 billion. In the months after January 1990, this agreed limit moved steadily towards the £8 billion judged to be unfinanceable at the start of the year. The refinancing package finally settled, later in 1990, at £8.7 billion, which involved Eurotunnel in raising an extra £2.7 billion. A further £2.1 billion in loans and £566 million in new equity through a rights issue (a lower gearing than the initial one) were secured late in 1990, and a further £100 million is anticipated from the proceeds of warrants. A summary of costs and financing arrangements is given in Table 3.1.

Agreement has been reached between Eurotunnel and TML on most, but not all, cost disputes: some elements are being referred to arbitration. As well as being greatly in excess of what had previously been thought a maximum financing limit, the new agreement is based on a lower Tunnel specification. Eurotunnel has been obliged to accept lower speeds through the tunnel for trains and shuttles, open rather than closed freight shuttles, a less glamorous Tunnel portal on the French side, and so on. The company was also obliged to make some changes in its senior management, with, for example, Alastair Morton moving from co-chairman to chief executive.

Two points can be made here. One is that all three parties — Eurotunnel, TML, and the banks — clearly had to engage in an element of brinkmanship, since each had vital interests at stake, and would gain or lose from the final settlement. However, each party was also concerned that brinkmanship should not provoke a terminal crisis, since each had more to lose from project failure than from an unsatisfactory (to it) agreement to continue. Paradoxically, the party which had least to lose from project failure was Eurotunnel, which had little to lose but its existence, and which might in any case find new contractors or new sources of finance. By contrast, the construction companies joined in the TML venture, and the lead banks in the banking syndicate, both had considerable prestige to lose from a failure to estimate costs correctly at the start of the project, and from a consequent failure to complete the project on time and to budget. A further party — the two governments — would also be embarrassed by project failure to the extent that they

Table 3.1 *Financial summary of Channel Tunnel project (1990)*

Costs[a]	£ million
Target works[b]	2,009
Lump-sum works[c]	1,305
Procurement items	583
Bonuses and contingency	311
Total construction	4,208
Corporate costs	787
Financing costs	1,386
Inflation	1,031
Net cash outflow	196
Total costs	7,608
Finance	
Initial equity and Equity I	47
Equity II (Oct. 1986)	206
Equity III (Nov. 1987)	770
Rights issue (Dec. 1990)	566
Total equity	1,589
Initial loan (Nov. 1987)	5,000
Additional loan (Nov. 1990)	1,800
New EIB facility (Nov. 1990)	300
Total loan finance	7,100
Proceeds of warrants less interest guarantees	23
Total finance	8,712

Source: Eurotunnel (1990b)
Notes: [a] Eurotunnel estimates of costs to completion, construction costs at 1985 prices.
[b] Works subject to a sharing of responsibility for cost overruns, principally tunnelling.
[c] Works subject to a fixed price — terminals, fixed equipment.

had taken the view both that private-sector financing and construction were possible, and that CTG-FM had an appropriate scheme and structure to effect it. Not only would the two governments' judgment in this case have been called into question by a bankrupt and failed tunnel, but also the ability of government to devise adequate procedures to make such judgments would have been doubted.

A second point to be made here is that the two governments had clearly created a problem for themselves. The no-guarantee position which they adopted meant that any financial assessment made by government could not be made public, since it would interfere with, and possibly distort, the market assessment. Moreover, the British government had placed itself in a position where it claimed not to be able to make any assessment of the Tunnel's non-commercial benefits, lest such an assessment should indirectly affect financing of the Tunnel. Thus, the government argued that if, for example, it had authorised an independent quantification of possible economic benefits to Britain, or to individual British regions, resulting from the construction of a Channel Tunnel, then Eurotunnel's commercial assessment of particular traffic would have been influenced. On this basis, it felt able to justify its 'do-nothing' policy stance, by which it declined to intervene either to capture any latent positive economic benefits associated with the Tunnel, or to diminish any of its evident economic costs. The regional consequences of this 'no-policy' policy stance could be substantial. Its most controversial product is section 42 of the Channel Tunnel Act 1987, which precludes public subsidy of through rail services to the Tunnel. This clause was inserted in the Act in order to convince port and ferry interests that they would be able to compete with the Channel Tunnel on a fair basis: the level playing-field argument. To the extent that this section has at the very least retarded the development of both major rail infrastructure improvements, and improved services to and from regions beyond London, it has affected the balance of traffic between shuttle and rail services through the Tunnel, and hence Eurotunnel's costs. Though initially not premissed on a dedicated UK rail link, these subsequently included this component (only to have it withdrawn again as the UK rail link ran into difficulties).

3.3 Traffic forecasting

Traffic forecasting depends on three basic factors: adequate information about existing traffic flows and their underlying patterns of demand; assumptions about factors likely to influence future growth; and assumptions about future market structure and competition. In the case of a transport project as large and complex as the Channel Tunnel, it can be argued that none of these three factors, and particularly the last two, can be satisfactorily supplied. Hence, Eurotunnel's revenue forecasts are always likely to be suspect, and continually subject to revision. This is not to criticise Eurotunnel, but simply to recognise that it is engaged in an impossible task: firm revenue forecasts cannot be made. The same problem naturally arises in the public sector, which consistently misforecasts both industries where fares are charged (and information is therefore more reliable) and industries where they are not. In both Britain and France, the traffic potential of new rail schemes has been consistently underestimated, and in both countries the traffic potential of new roads has been both under- and overestimated.

Existing flows are poorly documented for most international traffic, and particularly for traffic between the UK and continental Europe. Data exist for gross international flows by country, and for traffic flows over specific routes, but relatively little is known about the relationship between these revealed flows and the underlying demands. The advent of the Channel Tunnel has marginally improved the situation, since Eurotunnel (on payment of a fee) has persuaded the British government to improve its International Passenger Survey, and to update the International Freight Survey, which was last conducted in 1978. This has produced better data, not just on port to port flows but also on more complete origin–destination flows by mode and route. However, the use of even these data to a forecast of Channel Tunnel traffic depends crucially on assumptions about how the Tunnel will affect the existing market. French researchers have argued that understanding the impact of the TGV depends on viewing it as a new mode of transport, not just as a rail improvement (Laboratoire d'Economie des Transports/Interalp, 1986). The Tunnel is certainly a new mode of rail traffic, offering direct high-speed services between Britain and mainland Europe, but it could be argued that the shuttle offers only a marginally different service from a ferry.

The Tunnel involves a complex mix of rail and shuttle, each in competition with the other and, in differing degrees, with the alternative modes of air and ferry. It is not a complete transportation system in its own right, solving basic transport demands, but depends on connecting networks, more so possibly than airlines, and certainly more than the French TGV. This makes it extremely difficult to make a simple a priori assessment of its impact.

Transport demand, for both passengers and freight, is a function of need to move from one location to another. The demand for an individual mode of transport is a function partly of this, and partly of the supply characteristics of the individual mode, which determine its so-called generalised price. This price comprises not just monetary cost, but also time price, relative comfort, frequency, reliability, and so on. Hence, to forecast Channel Tunnel traffic it is necessary to relate existing traffic flows to existing characteristics both of the origin and destination economies, and of the quality of transport between them. Forecasting future transport demands then depends on forecasts both of the future of these economies and of future changes in the modal characteristics of transport. These latter forecasts must cover all modes, since both absolute and relative characteristics are relevant. There is, however, a further problem. The forecasts outlined here depend on traffic responding to different factors in a constant manner through time. Sometimes, however, modal changes can be so large as to upset the basic relationships which structure current traffic flows. This is likely to be the case for at least some cross-Channel traffic when the Channel Tunnel is opened.

All forecasts extrapolate existing trade and traffic flows in order to generate future demand functions. Table 3.2 illustrates some of the problems of this with respect to forecasts for previous projects (see

Table 3.2 *Cross-Channel traffic forecasts and growth (in percentages)*

	1962	1971	1973	1980	1990	1993	2000
Passengers in vehicles							
Actual	1.4	4.1	4.8	9.2	16.7		
1963 Forecast		2.2		2.6			
1971 Forecast				7.5	13.6		
1982 Forecast					17.9		25.5
1990 Forecast						20.5	27.3
Passengers not in vehicles							
Actual	2.9	4.8	5.2	9.0	6.0		
1963 Forecast		3.4		3.7			
1971 Forecast				7.0	9.7		
1982 Forecast					14.9		19.4
1990 Forecast						16.0	26.0
Air passengers							
Actual	5.3	8.9	19.6	28.0	39.0		
1963 Forecast							
1971 Forecast				23.7	41.5		
1990 Forecast						47.7	65.7

Sources and Notes
1963: Ministry of Transport (1963). No figures are given for air traffic, forecasts are only for tunnel traffic and not global market, assuming tunnel in service by 1969.
1971: Department of the Environment (1973a; 1973b; 1975), assuming tunnel in service by 1980.
1982: Department of Transport (1982), assuming tunnel in service by 1990, air forecasts were only made for London–Paris/Brussels routes.
1990: Eurotunnel (1990a); assumes tunnel in service by 1993, 1993 figures given as for full year, 2000 figures relate to forecasts for 2003; Eurotunnel forecasts are for global market and tunnel traffic only, these have been adjusted to allow for a continuing residual non-vehicle ferry traffic.

Vickerman and Flowerdew, 1990, for a fuller discussion). Typically, the sensitivity of traffic flows to variations in rates of economic growth is assessed, and forecasts are prepared for different growth scenarios. Hence Eurotunnel's revenue forecasts respond to different assumptions about the rate of economic growth in its core markets, compared with assumptions about transport changes, as shown in Table 3.3.

However, these forecasts are prepared on the basis of recent historical experience. Transport demand is strongly influenced by cyclical fluctuations in economic activity, but the strength of this relationship can change substantially. The period 1970–80 was, for example, characterised by a strong increase in British trade with the EC, and, within this pattern of overall growth, by a redistribution of British trade away from traditional partners like Belgium, the Netherlands and Portugal, and towards Italy and above all the Federal Republic of Germany. The realignment of the 1970s continued through the recession of the early 1980s, but at

Table 3.3 *Sensitivity of revenue forecasts*

	1993	2003	2013
Base case (2–2.5% GDP growth)	0	0	0
Low growth (1.4–1.75% GDP growth)	−7.3	−13.0	−17.7
High growth (2.4–3.0% GDP growth)	+5.0	+8.8	+13.2
With Channel Tunnel rail link	+1.6	+1.3	–

Source: Eurotunnel estimates

a slower rate. The 1990s should see the development of a different type of relationship, both as a result of completion of the single European market, with its impact on effective economic distances within the Community, and as a result of changing economic strengths. The emergence of Spain as the most dynamic European economy, and the consequences of opening up Eastern Europe to freer exchange, are major issues, the latter totally unforeseen.

Assumptions about future market structure and competition are thus difficult to make; indeed, a large part of this chapter is concerned with this issue. Initially, various competing claims were made. The British government saw a private-sector Tunnel as increasing competition in a market where collusion or regulation had kept fares among the highest per unit distance in the world. Until 1974 cross-Channel ferries had operated a revenue pooling cartel which was then outlawed by the Monopolies Commission (1974). This ruling led to some use of peak load pricing — hitherto charges had been standardised regardless of seasonal, weekly and diurnal fluctuations in traffic — but load factors remained very low on average, as operators sought to provide capacity for as much peak load traffic as possible. Attempts by the main ferry operators on short sea routes across the Channel to merge or to collude have so far been resisted by the UK Monopolies and Mergers Commission (1989b) on the grounds that they are against the public interest. In the air, binational regulatory agreements kept fares high, particularly in the lucrative London–Paris corridor, the densest international corridor in Europe.

The Tunnel will have a natural advantage over existing cross-Channel modes, because of its very low marginal operating costs and large capacity. Indeed, in principle the increase in capacity on cross-Channel routes which already have substantial excess capacity could pose a threat to the continued viability of ferry operations. However, it could also lead to a rationalisation of capacity. Furthermore, Eurotunnel, as a private-sector operator, is constrained in its pricing strategy. It will not be able to price at marginal cost until it has amortised the huge initial debt of constructing the Tunnel. Similarly, the ferries are unlikely to attempt to exploit this constraint by cutting their fares excessively, since bankrupting Eurotunnel would only have the effect of writing off the Tunnel's debt, and enabling a new operator to charge even lower fares. This would raise the problem of a monopoly Tunnel operator, but, even without regulatory interference by government, such a situation would be controlled by the relative ease of competitive entry into the ferry market.

However, it would be misleading to imply that the real competitive issue is simply characterised as Tunnel versus ferries versus airlines. Each offers differentiated as well as similar products. The ferry and airline industries are not each single entities, and there is scope for considerable competition within each sector between operators. Furthermore, ferry and airline companies operate on a variety of routes, unlike Eurotunnel, which will be a single-route operator. Each of these issues is discussed at greater length in the next two sections.

3.4 Pricing

In the final analysis, it is difficult to distinguish pricing and competition, and to investigate them separately. However, it is worth making an initial separation between them, in order to identify pricing issues which are internal to Eurotunnel, and those which are determined by outside competitive pressures.

The Channel Tunnel represents an interesting case of conflict between several pricing issues. It has already been noted that the purest case is that of the classic fixed-capacity, zero-marginal-cost crossing described by Dupuit (1844). This provides a clear rule about price as an allocative tool: at levels of demand below capacity, price should equal marginal cost (which is assumed to be zero, in that the marginal user imposes no discernable cost). Such a rule can, however, only be practically applied by a public utility, where consumer surplus is greater than total cost, and that cost is borne in full by the public sector.

However, such a rule is also applicable to the case of pure public goods, which are usually defined as non-rival in consumption and non-excludable. At levels of usage below capacity, the Channel Tunnel is clearly non-rival, but it is not non-excludable: like any fixed crossing, but unlike ordinary roads, it can quite easily be used to generate tolls for use. Indeed, it can be argued that the problem with a fixed link like the Channel Tunnel is that it has considerable monopoly power, which means that it can raise price well above marginal costs, and thereby effect a transfer of consumer surplus to itself as profit. This situation is depicted in Figure 3.1, where a monopoly Tunnel could raise price to P_m, involving a loss of P_mRSP_p in consumers' surplus, as compared with the public-utility situation of price equal to P_p.

The issue, then, is whether Eurotunnel's total profits, P_mRTP_p, are sufficient to amortise the Tunnel's debt, while at the same time the price P_m is not too high to provoke the two governments to perceive a need to step in and regulate pricing. Thus, the classic transport dilemma is identified — potential conflict between a price too low for profitable survival and a price too high for public acceptability.

Eurotunnel will, however, face a further pricing problem, that of allocating the Tunnel's fixed available capacity between alternative types

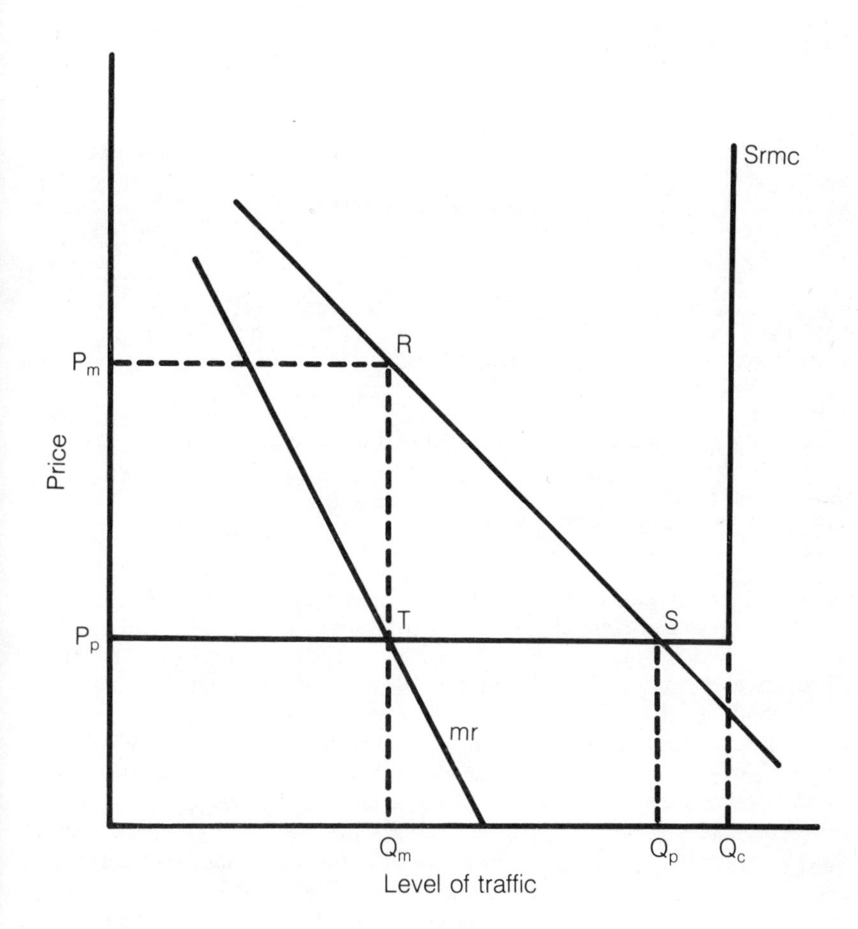

Figure 3.1 *Basic economic analysis of Tunnel pricing*

of traffic. At its simplest, shuttle trains and through rail services can be compared. Eurotunnel has concluded an agreement with BR and SNCF which lets 50 per cent of the Tunnel's capacity to the rail companies in return for a relatively large fixed sum and much smaller fees per unit of traffic. From Eurotunnel's point of view, this has the benefit of guaranteed revenue, though through rail services are also a competitor to Eurotunnel's own shuttle service. Therefore, the demand curve for the shuttle is shifted, and Eurotunnel's pricing freedom is reduced.

There are, however, two reasons why it would not be in Eurotunnel's interest to attract too much traffic away from rail. First, every unit attracted from rail adds to potential congestion in the Tunnel. Eurotunnel, therefore, needs to keep as close as possible to the rough 50–50 split of traffic forecast in order to avoid distortions in capacity utilisation.

Table 3.4 *Illustrative costs for 38 ton lorry and driver on selected UK-Continental journeys, 1989*

	Road distance (km)[a] UK	Continent	Total	Cost of lorry plus driver (£)	Ferry cost (£)	Ferry cost as % of total cost	% saving using Tunnel[b]
London–Cologne	125	409	534	272	198	42	4.3
Manchester–Marseille	450	1,059	1,509	682	198	23	2.3
Glasgow–Rome	780	1,662	2,442	1,177	198	14	1.4

Notes: [a] Via Dover–Calais
[b] Based on assumption that the Tunnel is 10 per cent cheaper than Dover–Calais ferry.
Source: Pieda (1989).

Second, through rail traffic is much less susceptible to competition from ferries than is Eurotunnel's shuttle traffic, which can switch at the last moment. Congestion in the Tunnel is therefore a greater threat to Eurotunnel than to the rail companies, and would benefit the ferries.

On pricing, therefore, Eurotunnel would appear to be caught in a difficult situation, where freedom to exploit monopoly power is severely limited (Kay *et al.*, 1990). There are, however, further problems posed by the demand curve it faces. Typically, demand for transport is strongly price-inelastic. Therefore, price reductions are rarely an effective way of increasing total traffic, and certainly not of increasing revenue. Thus, even if Eurotunnel were to cut cross-Channel fares substantially, it would be unlikely to increase the volume of cross-Channel traffic, even if cost savings were reflected in the total cost of journeys.

Usually, however, the Channel crossing is only a relatively small element in most journey costs, ranging from perhaps 40 per cent of a journey from London to Cologne, to 15 per cent or less of a journey from Glasgow to Rome, as shown in Table 3.4. If demand were really price-inelastic, then cross-Channel operators might be encouraged to feel that they could raise prices without loss. Recent evidence of passenger demand for BR's services (Owen and Phillips, 1987) suggests, however, that the elasticity is near to -1.0, which would imply no revenue enhancement from raised prices. It is not clear that the composition of passenger traffic is sufficiently similar to justify carrying this estimate over to the cross-Channel market. However, there is certainly no suggestion in these figures that, independently of the competitive situation and of its share of the market, Eurotunnel can actually substantially change the size of the overall market through its pricing policy.

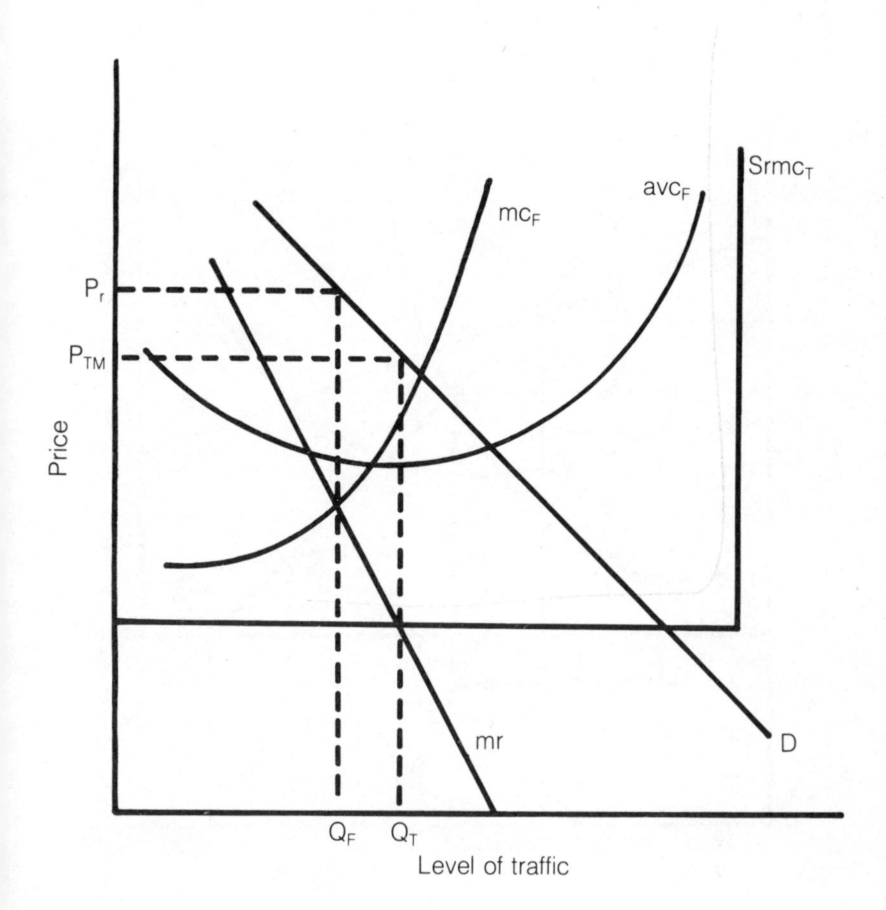

Figure 3.2 *A simple competitive model*

3.5 Competition

Competition is perhaps the central concern of this chapter. How does the Channel Tunnel fit into the market for cross-Channel transport? What will be its impact on prices and on profitability? It has been suggested that the scope for using absolute price levels as a means of increasing the total cross-Channel market is likely to be limited. However, relative price may be a major factor in determining the market shares of different operators.

We start by returning to our simple diagrammatic model of Figure 3.1, and add to it the position of the ferries. For simplicity, we assume, as in Figure 3.2, that ferry companies face a U-shaped variable cost curve, avc_F, which, at all levels of traffic, is above the marginal costs of the

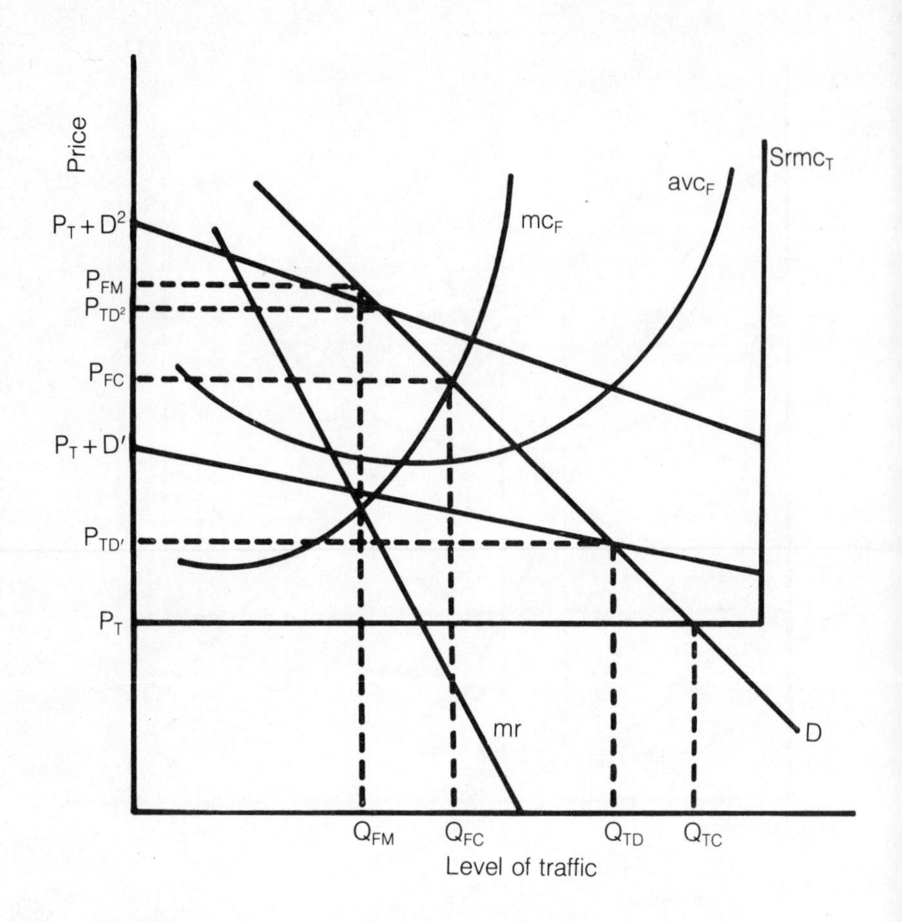

Figure 3.3 *Competition allowing for debt burden*

Tunnel. In these circumstances, the ferry companies would fix their joint profit-maximising price at P_F, where marginal revenue equals marginal cost, mc_F. On this basis, the Tunnel could beat the ferries on price because of its lower marginal costs, and could capture the entire market because of its large capacity.

However, the situation in Figure 3.2 is oversimplified for three basic reasons. First, Eurotunnel will be unable to price at marginal cost because of its large debt burden. Second, ferry companies are prohibited from colluding, and, therefore, cannot operate to maximise joint profits. Third, cross-Channel traffic is not homogeneous, and, therefore, all categories do not display similar price elasticities of demand, nor do they incur identical marginal costs. In addition, daily and seasonal peaks imply problems in obtaining efficient utilisation of capacity for both Tunnel and ferry operators.

Figure 3.3 demonstrates two alternative assumptions about the size of the debt burden. If this is fixed, then its average cost will fall as usage increases. However, this may imply an average cost either above or below the ferries' average variable cost. If below that of the ferries, at $P_T + D^1$, then Eurotunnel is still market leader, though with a lower level of usage. If above that of the ferries, at $P_T + D^2$, then the ferries can beat Eurotunnel on price by charging at or slightly above their marginal cost price, $P_F C$. The question then is whether they have the capacity to sustain this price, and indeed whether in the long term it is in their interests to do so.

Two issues are relevant here. We have taken the ferries' variable costs as the main determinant of competitive position in the short term, on the assumption that they will have a substantially amortised fleet in 1993, and that capital charges will be of greater relevance to Eurotunnel. However, in the longer term the ferries will need to make provision for depreciation and reinvestment to a far greater extent than will Eurotunnel. To dominate the cross-Channel market, therefore, the ferries need to bankrupt Eurotunnel. This, however, raises the subsidiary point, already noted, that by doing this the ferries actually eliminate their advantage, since a bankrupt Eurotunnel will face a substantially smaller debt burden. Hence it is not in the ferries' interests to be excessively aggressive in their pricing policy.

Thus far, the model has assumed a clear, dominant market leader, who can effectively fix a price which captures the entire market. Figure 3.4 attempts to depict a more likely compromise situation in which, given its short-term marginal cost advantage, Eurotunnel is price leader, but does not seek to destabilise the market by exploiting its advantage to the full. It therefore settles for an acceptable share of the market, at Q_T, and prices accordingly, at P_T, leaving the ferry companies facing not demand curve DD^1, but a residual demand curve $P_T D_F$. The ferries' survival then depends on their ability to reduce costs below avc_F to secure a share of the market. In this situation, Eurotunnel is able to set a price below P_T, which allows it to control the amount of ferry operation which would best suit its own objectives.

At this stage we also need to introduce the structure of the ferry industry itself. Since the UK Monopolies Commission ruling in 1974, the ferry companies have not been allowed to collude and operate an effective monopoly through a joint-ticketing revenue policy, and collaborative scheduling. This ruling has recently been confirmed by the UK Monopolies and Mergers Commission (1989b). Hence we need to make some assumptions about the response of ferry companies to each other as well as to the Tunnel. Again, we might use a leader-follower assumption that one company occupies a dominant position, fixing price and capacity such that the others take the residual market. Perhaps more appropriate, however, is an assumption of duopoly, with a number of fringe operators exploiting particular niches in the market. Certainly there is little evidence of real price competition between the major ferry operators on the short sea routes — Sealink, jointly operated by British

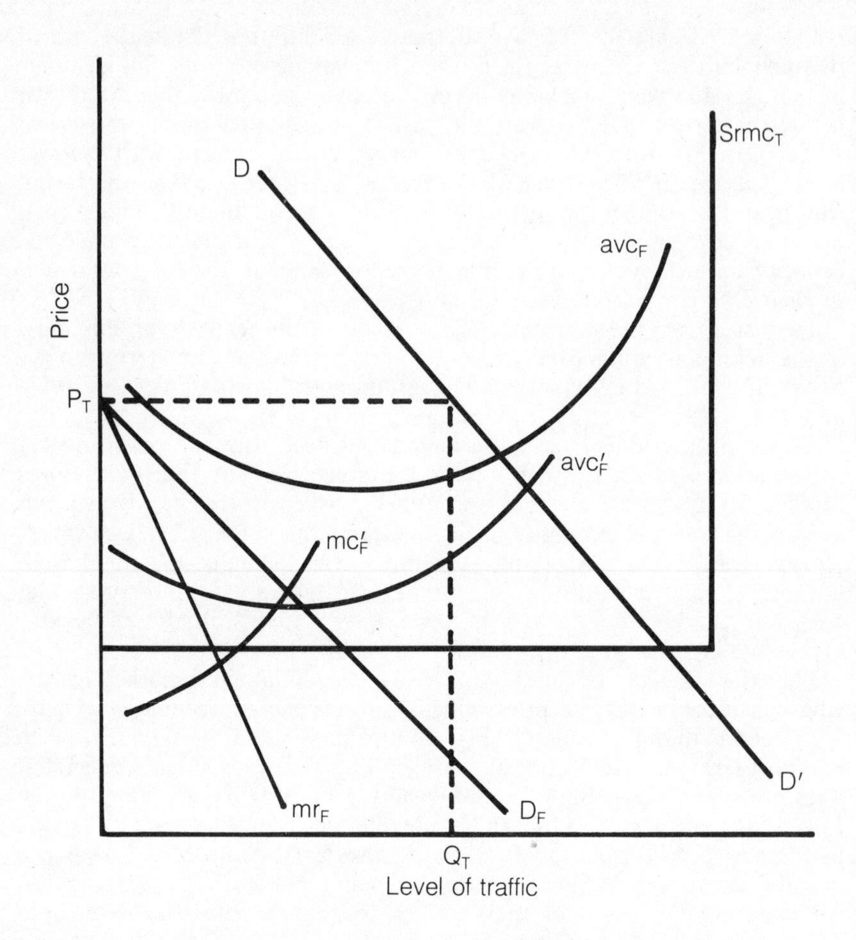

Figure 3.4 *Price leadership in the cross-Channel market*

Ferries (as successor to British Rail) and SNCF, and P&O European Ferries (which operates jointly with RTM on the Dover–Ostend route). Although timetables are not formally co-ordinated, they tend not to provide direct competition at specific sailing times, partly because of constraints on port capacity and operation. Hence competition has been in terms of quality of on-board services (including duty-free shopping), regular traveller discounts and the marginal time advantage which P&O's new faster ships enable it to offer. Interestingly, there is little evidence that its 15-minutes faster crossing has resulted in a major competitive advantage to P&O.

The relative stability resulting from this duopoly in the market can also be seen in the poor capacity utilisation of the ferry fleet. Since average cost curves are likely to display a rather flatter U-shape than those

Table 3.5 *Variations in traffic, profits and prices with market structure*

	Number of ferry companies in the market			
	1	2	3	4
Traffic of each ferry company (mn units)[1]	53.3	40.0	32.0	26.6
Traffic of the Tunnel (mn units)[1]	72.1	58.8	50.8	45.4
Profit of each ferry company (£mn)	270.8	146.5	89.0	57.7
Profit of the Tunnel (£mn)	244.7	70.3	−17.2	−68.5
Price per unit (£)	7.46	6.13	5.33	4.79
Total traffic (mn units)[1]	125.4	138.8	146.8	151.8

Note: [1] Traffic is expressed in terms of units where foot passengers count as 1 unit, bus passengers = 1.5, car passengers = 3 and freight (tonnes) = 2
Source: Duchêne (1988).

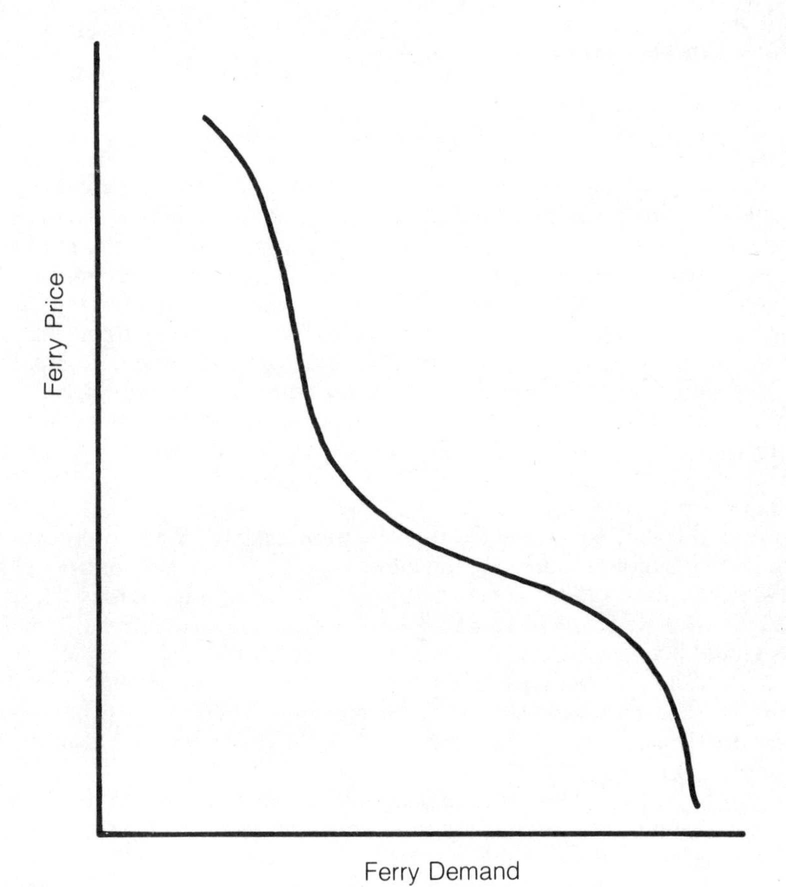

Figure 3.5 *Stable and unstable equilibria in the ferry market*

Table 3.6 *Prices and market shares*

Prices (£, 1994 prices)	Pre-Tunnel	Post-Tunnel Dover ferry	Post-Tunnel Non-Dover ferry	Post-Tunnel Tunnel
Foot passengers	10.87	8.23	13.11	7.65
Cars	115.42	50.72	85.16	41.44
Freight	214.16	177.04	196.19	128.79
Air diverted passengers	75.54	–	–	46.00

Market share	Dover ferry	Non-Dover ferry
Foot passengers	0.13	0.17
Cars	0.12	0.21
Freight	0.04	0.27
Air diverted passengers	–	–

Source: Kay *et al.* (1989).

assumed in earlier models, there is little incentive for the companies to remove spare capacity, which is useful for meeting peaks. At the same time there is no real incentive to capture a larger share of the market on cost grounds. However, the entry of the Tunnel into the market does change this situation, since a similar proportional distribution of reduced, post-Tunnel ferry traffic would very probably force the ferry companies on to a rising part of their average cost curves. Competition will then be intensified between the ferry companies where, given similar cost structures, there can be winners and losers, rather than between any one ferry company and the Tunnel. Indeed, if a single ferry company took on Eurotunnel it would be likely to lose out principally to the other major ferry company.

It is difficult to reproduce this situation exactly in an empirical way, but some simple numerical simulations have been carried out (Duchêne, 1988; Dormard, 1988). These begin by translating the various classes of traffic into standard units based on costs as reflected in relative tariffs. A global demand function is then derived for 1980 traffic on the assumption of linearity with price elasticity of unity at the ruling price level. Using basic cost information from the 1982 UK/French Study Group Report (Department of Transport, 1982), cost functions are then derived for each mode.

On the basis of these sets of information, a simulation of the situation under assumptions of different numbers of ferry companies has been undertaken, as shown in Table 3.5. It was then used to derive separate demand functions for airlines and ferry companies. The global demand function was then projected forward from 1980 to 2000. The major result of this simulation is to confirm that although ferry companies can

Table 3.7 *Welfare analysis (£ millions, 1994 values)*

	Net social benefit	Private benefit to Eurotunnel
Consumers	31,039	–
Ferry operators	– 1,577	–
Tunnel operators	18,336	15,942
Air/rail operators	– 4,143	–
Cost of building Tunnel	– 11,839	– 11,839
Total	31,816	4,103

Source: Kay *et al.* (1989).

make severe inroads into Channel Tunnel profitability by increasing capacity and reducing fares, the impact of extra ferry competition is felt very clearly by the ferry companies themselves.

Using a more sophisticated formulation, Craven (1988) also reproduces possible Tunnel–ferry competitive scenarios. This model allows for possible congestion building up at certain times on the Tunnel, and for peak load pricing to ration demand at this point. This model implies non-linear demand curves for both ferry and Tunnel modes, with a pronounced S-shape (Figure 3.5) reflecting different responses to price changes. The competitive response of both operators is taken as Cournot; that is, each assumes that the price–output decisions of its rival will not be altered by its own actions. Under slightly different conditions, this model can produce both stable and unstable equilibria as posited originally. The capacity limit of each operator emerges as critical in this model.

Further work in this area has been carried out by another group of researchers (Kay *et al.*, 1989; 1990). Their model incorporates an explicit user-preference element, so that a measure of welfare gain as well as operator profit can be obtained. In the passenger market, ferries are assumed to operate as a collusive monopoly, but some competition is allowed in the freight sector. Ferries are assumed to be profit maximisers with Cournot expectations.

An important element in this model is its attempt to include the greater variety of routes operated by ferries, which provide a form of competition to which the Tunnel, with its fixed routing, cannot respond.

The base results of this model are reproduced in Table 3.6. Noteworthy is the substantial price reduction which the model generates, with the Tunnel picking up very substantial market shares. For example, car prices are reduced by 64 per cent, with the Tunnel gaining 88 per cent of the market. In this situation, the Tunnel is both commercially and (even more) socially viable, since the massive price reductions produce a substantial consumer surplus. Table 3.7 reproduces these results over the full concession period.

Kay *et al.* (1989) also investigate the consequences of varying certain

Table 3.8 *Comparison of rates of return (per cent)*

	Financial return		Cost–benefit analysis	
	Low growth	High growth	Low growth	High growth
1963 Study[a]	6.05	8.25	10.4	13.3
1973 Study[b]	14	17	14.6	17.6
1979 Study[c]	12.6	14.9	15.7	19.2
1982 Study[d]	–	–	4.7	9.3
1987 Projections[e]	16.6	18.8	–	–
1990 Projections[f]	14.1	16.9	–	–

Sources & Notes:

[a] Ministry of Transport (1963). The financial analysis is not based on a full net present value calculation, but on the ratio of estimated dividends to share capital in years 13–17.

[b] Department of the Environment (1973a; 1973b).

[c] Commission of the European Communities (1980), and see Gwilliam (1983).

[d] Department of Transport (1982). Uses a different basis with much more rigorous assessment of resource cost implications for existing modes.

[e] Eurotunnel (1987). Gross dividend yield over the life of the project, the upper and lower values are based on assumptions of 10% increase or decrease in revenues around base case.

[f] Eurotunnel (1990b), definition as for 1987.

parameters, such as a lower rate of growth for a substantial reduction of ferry marginal costs. These can both affect Eurotunnel's profitability substantially, but do not diminish social returns.

However, two issues — the validity of the competitive response assumption, and the question of market segmentation — require further consideration. The Cournot hypothesis is convenient as a working hypothesis to establish an economic framework. Conventional models seem much less appropriate in this case than in most, however, because of the complexities raised by the different cost structures of the operators. Where firms know that their competitors face similar cost structures to themselves, they may be less interested in looking at the implications of dynamic responses. Although adopting a short-run Cournot strategy may mean getting things wrong in a typical competitive situation, the costs of being wrong are both fairly well known and manageable. In the case of competition with Eurotunnel, the costs to the ferries of being wrong are both unknown and quite probably catastrophic. This suggests that competitors will be much more cautious before launching into a full-scale price war. Eurotunnel, as the new entrant, has much better information about its competitors than the latter have about it. It can use this to place itself in the position of market leader. This makes analysis of market outcomes much more difficult. However, it does seem likely that the general level of prices will fall by much less than Kay *et al.*'s base case.

Secondly, the question of market segmentation is important. Modes of transport serving the same route can coexist because they offer different services which appeal to different types of user. Hence the market becomes segmented according to particular demands for speed, price, frequency, comfort, and so on. By gearing provision to one particular segment, an operator can gain greater pricing freedom by effectively creating a monopoly for the provision of the differentiated service. However, it is not immediately clear that Eurotunnel does offer a clearly differentiated service. In terms of overall journey times, there is only a marginal gain for most shuttle users, though a potentially substantial gain exists for through rail users. Service reliability could, however, be greater for the Tunnel, which is less affected by weather conditions. By reducing the variance in journey times, this could command a premium for the Tunnel for certain classes of traffic.

Market segmentation implies the possibility of price discrimination as a means of increasing revenues. However, in the case of cross-Channel traffic, segmenting the market can also lead to variations in the marginal cost of carrying each segment. One particular problem is the cost imposed by variations in traffic flow through time. Any operator which can achieve a greater capacity utilisation will be able to earn a greater return. Real competition is, therefore, for base-load traffic, and a successful strategy involves shifting traffic which only occurs in peak hours and days to competitors who need to maintain expensive capacity to cater for it.

3.6 Overall investment appraisal

This chapter has investigated the basic cost and revenue elements facing Eurotunnel, taking into account its competitive position in the cross-Channel market. Above all, the chapter has demonstrated the complexity and uncertainty surrounding any serious investment appraisal exercise, even without taking social costs and benefits into account. Studies of previous Channel Tunnel projects did derive broadly consistent financial rates of return, of between roughly 12 per cent and 15 per cent, depending mainly on economic growth forecasts. Typically, these produced social rates of return on the basis of a transport cost–benefit analysis — that is, one which excludes wider economic impacts — of 14–18 per cent, as shown in Table 3.8.

By these criteria, the Eurotunnel prospectus and some of the early enthusiasts, who forecast financial rates of return of well over 20 per cent, must be seen as too optimistic. Such rates of return were based on traffic and revenue forecasts which accelerated fairly rapidly, only to be caught up and overtaken by costs during 1989. Probably the rate of return is now again in the historically typical range of 12–15 per cent, certainly in the medium term. Nevertheless, many still believe that for investors prepared to wait, Eurotunnel will eventually become a healthy investment. This perhaps calls into question the appropriateness of a conventional rate of return analysis for a project of this type, where an enormous capital outlay is incurred well before any revenues are earned.

Does this, therefore, make financing through a mixture of equity and conventional loans inappropriate?

This question of financing has been raised by a number of commentators, including Alastair Morton of Eurotunnel itself (cf. Gerardin, 1990; Morton, 1989). It returns us to the essential question of whether government (perhaps in an international form, like the European Commission) should be involved to effect this, not as a bottomless well of finance to support spiralling costs but as the only type of investor able to take a sufficiently long-term view of benefits. To some extent, the European Investment Bank has assumed this role as the largest single supplier of finance to Eurotunnel, albeit principally secured against finance from private-sector financial institutions. It is interesting to speculate that the result of the British government's reluctance to provide this bridging finance could have forced it to take the Channel Tunnel into public ownership through a rescue operation, when such initial finance could have led to a successful subsequent privatisation. Now, however, this is probably a strictly academic point, since both finance and the Tunnel's future seem secure.

Notes

1. External costs and benefits are considered in later chapters, notably Chapter 10.
2 Further, in April 1991, it was announced that the IGC was requiring modifications to the design of shuttle vehicles after delivery had commenced, which could delay the introduction of a full shuttle service for some six months after opening.

Part Two

Public policy responses

4 *The Channel Tunnel in economic context*

Analysis of the Channel Tunnel and public policy in Britain and France was a major concern of Part One. However, Part One focused almost exclusively on the internal dynamics of the present Tunnel project, and did little to situate the Tunnel in its external environment. In that environment, it will have significant impacts in both the short and long terms, and in both an immediate and an extended hinterland. The purpose of Part Two is to set the Tunnel in its external economic environment (Chapter 4), to present the range of institutions and actors which have been prime movers in devising policy responses to it (Chapter 5), to analyse those policy responses (Chapters 6 and 7), and to consider their appropriateness (Chapter 8).

At the outset, it should be noted that the divide between the internal dynamics of the present Tunnel project and its external environment is not absolute, and is in some ways artificial. In France, in particular, the fixed link has always been seen as but one link in a transport chain which will eventually stretch into many parts of Europe, and which will have a significant regional impact far beyond the Channel Tunnel itself. However, for purposes of analysis, these two aspects of the Tunnel project may be separated, although it should be borne in mind throughout Part Two that conclusions reached in Part One remain valid here, and are often relevant to discussion.

Thus, in Part Two, the Tunnel will be used as a tool for comparative analysis of public policy in Britain and France, the idea being that it sets similar challenges in the two countries, and thereby offers a means of comparing and assessing the policy responses of their systems of government. However, it should be noted that although the challenges set by the Tunnel in Britain and France may often be similar, they are never identical. Mere consideration of the economic situation of its immediate hinterland is sufficient to demonstrate this: on the French side, a region of heavy industry in structural economic decline since the early 1960s, seeking to reposition itself in the Europe without frontiers which will be created by the EC's 1992 programme; on the British, a rural and suburban county which has never been heavily industrialised, and which, although faced with economic problems in the 1980s, is chiefly concerned to retain its character as the 'Garden of England'.

In these circumstances, the first task of comparative analysis is to investigate the Tunnel's economic context, as a means of setting analysis on a firm footing: in a sense, it is to generate a 'level playing field' for comparison of public policy in Britain and France (or, more accurately, to show the places in which that playing field is uneven). The second task is to present the main political and institutional features of the British and French systems of government, in order to set the parameters of their capacities and capabilities. Only when these twin tasks have been accomplished can valid comparative analysis proceed.

The central argument of this chapter, and to some extent of this book, is that there is a basic asymmetry between the Channel Tunnel's economic situation in Britain and France, and that this is itself a major determinant of divergent policy responses to it. In this chapter, this asymmetry will be considered not only regionally, but also nationally and internationally.

4.1 France

In France, the Channel Tunnel emerges at Sangatte, just to the west of Calais, and part of the Nord-Pas de Calais region. This region is in the North of France, and shares borders with Belgium to the north and east, and with the French region of Picardie to the south. At its south-eastern tip lies the French region of Champagne-Ardenne. Nord-Pas de Calais is shown in Figure 4.1.

Traditionally, Nord-Pas de Calais is a bastion of Catholicism and of the French Left, which retains a dominant presence in the region, and is currently led by Pierre Mauroy, First Secretary of the Parti Socialiste (PS), Mayor of Lille, president of the Communauté Urbaine de Lille (and Mitterrand's first Prime Minister between 1981 and 1984). There are, however, important exceptions to the present predominance of PS control in the region. Boulogne-sur-Mer, Douai and Valenciennes all voted right wing at the last mayoral elections (in 1989); Roubaix voted centrist; and Calais retained faith with its long-serving Communist mayor.

Nord-Pas de Calais is small but populous: in 1990, 3.96 million people (7 per cent of the national population) occupied its 13 000 km^2 (2.3 per cent of the national territory). The main characteristics of this population were that it was young, concentrated and urban. In 1986, 33 per cent of Nord-Pas de Calais residents were under 20 (as against 27.5 per cent in France as a whole), and 16 per cent were over 60 (as against 18 per cent). The number of inhabitants per square kilometre was 316, as against 101 in France as a whole, and 104 in the EC. Eighty-six per cent of the population lived in cities, and the region counted eight metropolitan areas of more than 100 000 inhabitants each, putting them among the 80 largest urban areas in France. The eight were Lille–Roubaix–Tourcoing, Lens, Béthune, Dunkerque, Calais, Valenciennes, Douai and Maubeuge (Chambre Régionale de Commerce et d'Industrie Nord-Pas de Calais, 1989).

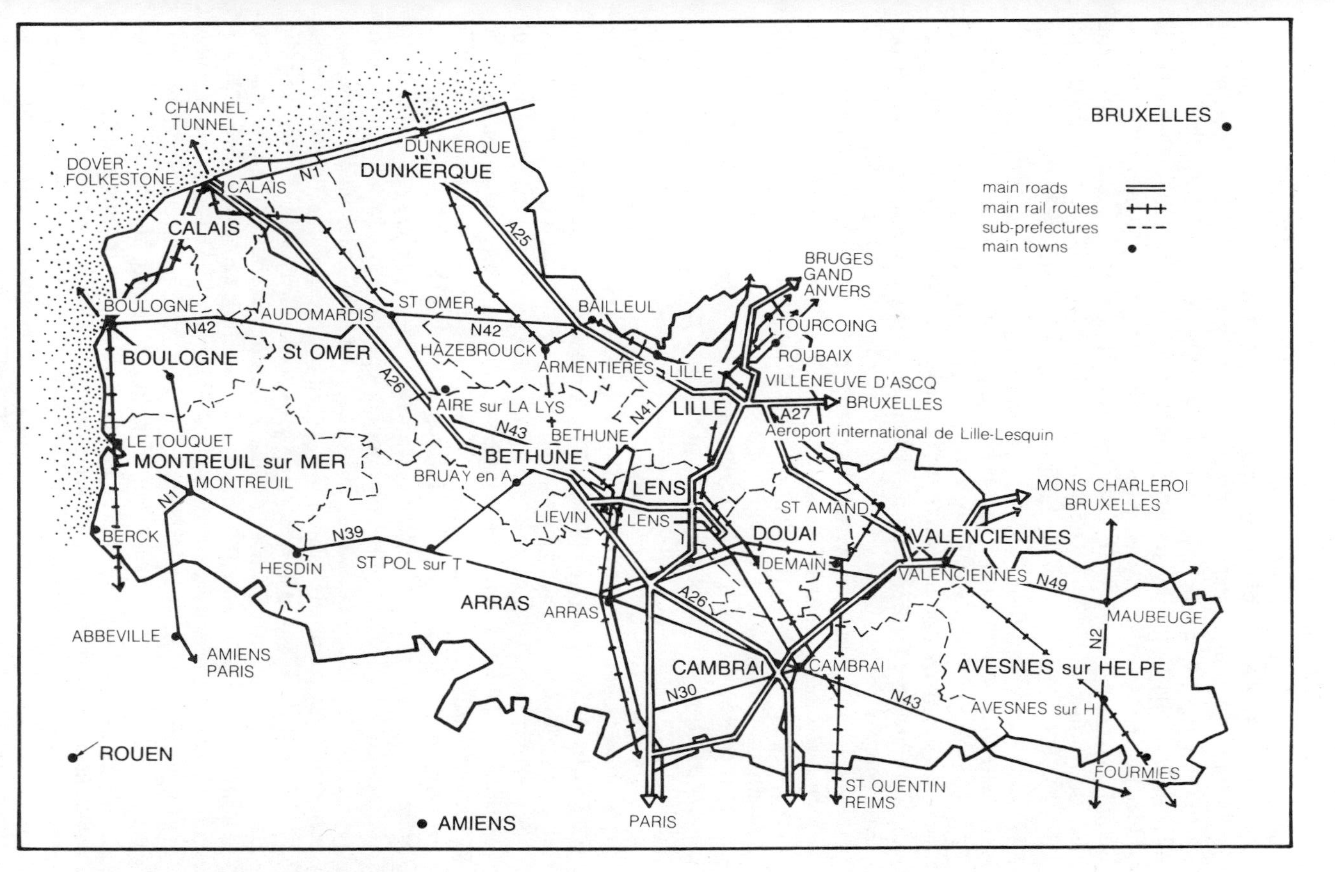

Figure 4.1 *Situation map: Nord-Pas de Calais*

A driving force of the French economy in the nineteenth century, with its 'smokestack' textile, coal and steel industries — which form the gloomy backdrop to Zola's critique of modern industry in *Germinal* — Nord-Pas de Calais, like many similar regions across Europe, is now in structural economic decline. In the postwar era, it was first hit by economic problems when its textile industry moved into crisis in the early 1960s. At this time, the industry employed just under 160 000 workers. One in ten of these was Belgian, but in 1962 textiles nevertheless employed one Nord-Pas de Calais worker in six. Subsequent contraction of the industry — in particular since 1967 — has been uniform and dramatic, except in years of high national consumption (1969, 1976 and 1979). Twenty-one thousand jobs were lost in textiles between 1968 and 1975; 41 700 between 1975 and 1982; and 11 000 between 1982 and 1985. As a result, by 1985 total employment had been halved to 80 000, and was projected to fall below 70 000 by the end of the decade. Although the proportion of foreign workers had been cut to between 2 and 3 per cent, only one regional worker in 14 was employed in textiles in 1985 (INSEE, 1986a). The provisions of the EC's Multifibre Arrangement, which provides non-tariff protection against imports from major low-wage importers, would seem to have done little to stem the decline (Shepherd, 1984).

At the same time as textiles was moving into crisis, the 1960 Plan Jeanneney outlined a phased reduction of regional coal production (Thiébault, 1986). Total regional employment in the energy sector, which had been over 200 000 after the War, declined from 103 500 in 1968 to 74 600 a mere four years later, to 58 100 one year later still, and eventually to 32 300 by 1985. Thus, 26 000 jobs were lost in this sector between 1975 and 1985: 28 000 were lost in coalmining, and 1000 in oil; while 3000 new jobs were created by Électricité de France-Gaz de France (EDF-GDF), half of these at the Gravelines nuclear power station, along the coast from the Channel Tunnel, and the largest such installation in Europe (INSEE, 1986a). In December 1990, coalmining ceased in Nord-Pas de Calais.

In the mid- to late 1960s, then, two of the region's three staple industries were faced with substantial contraction. In an attempt to revive the flagging regional economy, the French government sponsored ambitious restructuring plans, which, on occasion, conflicted with EC policy (Lenel, 1988). The centrepiece of its strategy was to persuade the then privately owned giant steel-maker Usinor substantially to increase its investment in the region (by means of huge injections of public funds),[1] and to constitute its motor of renewed economic growth. Around this strategy, centred on Usinor's main factories in Dunkerque, vast development plans were conceived, both on the northern Nord-Pas de Calais coast of Calais–Gravelines–Dunkerque and in the mining basin around Lens and Valenciennes. However, steel was also hit by crisis following the first oil-price explosion in 1973, and in 1977 the industry moved into precipitate collapse. In the next eight years, more than 20 000 jobs were lost, as regional employment in steel fell from around 50 000 in 1977 to just over 29 000 in 1985 (INSEE, 1986a).

Thus, the period since the early 1960s has seen the Nord-Pas de Calais region hit by a succession of crises, each of which has contributed to very high rates of regional unemployment. As total industrial employment fell from some 580 000 (in a regional workforce of between 1.1 and 1.2 million) in the early and mid-1960s, to 380 000 in the mid- to late 1980s, regional unemployment rose to 13.3 per cent at the start of 1986. This global figure masked rates of between 16 and 18 per cent in Calais, Boulogne, Valenciennes and the Avesnes region, and high variations between age groups (INSEE, 1986b). At the end of December 1987, unemployment in areas most affected by recession — in particular Calais and Boulogne — exceeded 20 per cent, and was highest among people under 25. On the coastal strip, youth unemployment averaged 37.2 per cent (Kent County Council/Conseil Régional Nord-Pas de Calais, 1988).

Responses to this crisis were various.[2] On the part of workers, massive job losses, particularly at Usinor, provoked strikes and demonstrations. On the part of politicians and administrators, these in turn provoked conciliatory reactions. Thus, violence in Denain and Valenciennes (and in Longwy, Lorraine) in 1979 persuaded the Socialists to adopt a policy of no compulsory redundancies which leading Nordiste Pierre Mauroy, demonstrator at Denain in 1979, simply could not abandon as Prime Minister between 1981 and 1984. The most optimistic scenario of the 1982 Judet Report, 'The trends in the steel industry's markets and its medium-term prospects', was adopted as the national target for steel output, and was only finally abandoned when austerity measures were imposed across the board by the Mauroy government in 1983 in the face of declining international confidence in, and support for, the franc. Losses at Usinor and the Lorraine-based steel-maker, Sacilor, continued after nationalisation by the Socialists as they had before: a combined 6 billion francs in 1981, over 5 billion in 1982, and 11 billion in 1983. At a press conference on 4 April 1984, President Mitterrand outlined a 'final' solution involving major industrial restructuring and redeployment, but offering no clear resolution of the economic problems of Nord-Pas de Calais.

In these circumstances, a 'retreat in good order' was organised by politicians and administrators at both national and local levels: often, of course, they were one and the same person. In the central regional conurbation of Lille, plans to develop the north–south axis of Lille–Roubaix–Tourcoing came into conflict with the pet project of Arthur Notebart, until 1989 President of the Communauté Urbaine, which was to promote an east–west axis through construction of a new town to the east of Lille. Conceived in the mid-1960s, Villeneuve d'Ascq, as it was later to be named, was to a large extent linked to the 1970s Channel Tunnel project, the hope being that east–west traffic through the Tunnel would stimulate high-technology development on the model of Boston's Route 128. This was not to be, and although huge sums were invested in the world's first automatic metro system, the VAL (developed by researchers at the Université de Lille I), the new *technopole* did not develop a high-technology image, and attracted only low-technology office and small industrial users.

Thus, by the beginning of the 1980s, when the idea of a Channel fixed link again returned to the policy agenda, the prospects of a region premissed on growth — whether realised, as in the long period of industrial preeminence stretching into the 1960s, or largely unrealised, as during the mounting economic crisis of subsequent years — were decidedly bleak. Visionary plans to develop the port of Dunkerque had been abandoned when Usinor collapsed in the mid-1970s, and the town was dealt a further decisive blow when the state-supported Normed shipyard was served with closure notices in 1984. Today, extensive areas of land zoned in the early 1970s for industrial use by state planners in the town's sub-prefecture (Agence d'Urbanisme et d'Aménagement de la Région Dunkerquoise, 1973) remain derelict. Equally ambitious plans to develop Villeneuve d'Ascq came to little when the 1970s Tunnel project was halted.

To such a region, the mere construction activity involved in building a Channel fixed link, never mind the chance in the link's operational phase to reposition the regional economy, was almost universally welcome. Pockets of resistance emerged — particularly in Calais, where the ferry industry was by a large margin the major employer, and in other port towns such as Boulogne and, to a much lesser extent, Dunkerque — but they were essentially marginal, and easily marginalised. By and large, the region quickly recognised the benefits it might draw from a fixed link, and as early as February 1982 — four years before the concession agreement was signed — the Regional Council had signalled its desire to see the project advance, and had stated the terms on which it would be of greatest benefit to the region.

4.2 Britain

In Britain, by contrast, the Channel Tunnel emerges at Cheriton, Kent, to the west of Folkestone, in a county which has suffered successive economic shocks since 1973, but which, even in the early 1980s, was not seeking, and probably could not have handled, major industrial investment and restructuring.

Kent is a large so-called shire county, with an estimated population in 1988 of 1.52 million, and a land coverage of 1440 square miles. More than a quarter of its people live in small towns and villages, but the main concentration of population is in large towns and urban areas. However, the county has no major town, the largest urban area being the Medway towns (Chatham–Gillingham–Rochester), with a combined population of 225 000. The population of Kent is significantly older than that of Nord-Pas de Calais: in 1981, 18.5 per cent of Kent's population was retired, as compared with 11.7 per cent of the 1982 population of Nord-Pas de Calais (Kent County Council, 1988; 1989). Politically, Kent is predominantly Conservative: the County Council, for example, has been under tacit or overt Conservative control since its institution in 1889, and few district councils or Parliamentary seats fall into Labour or centrist hands.

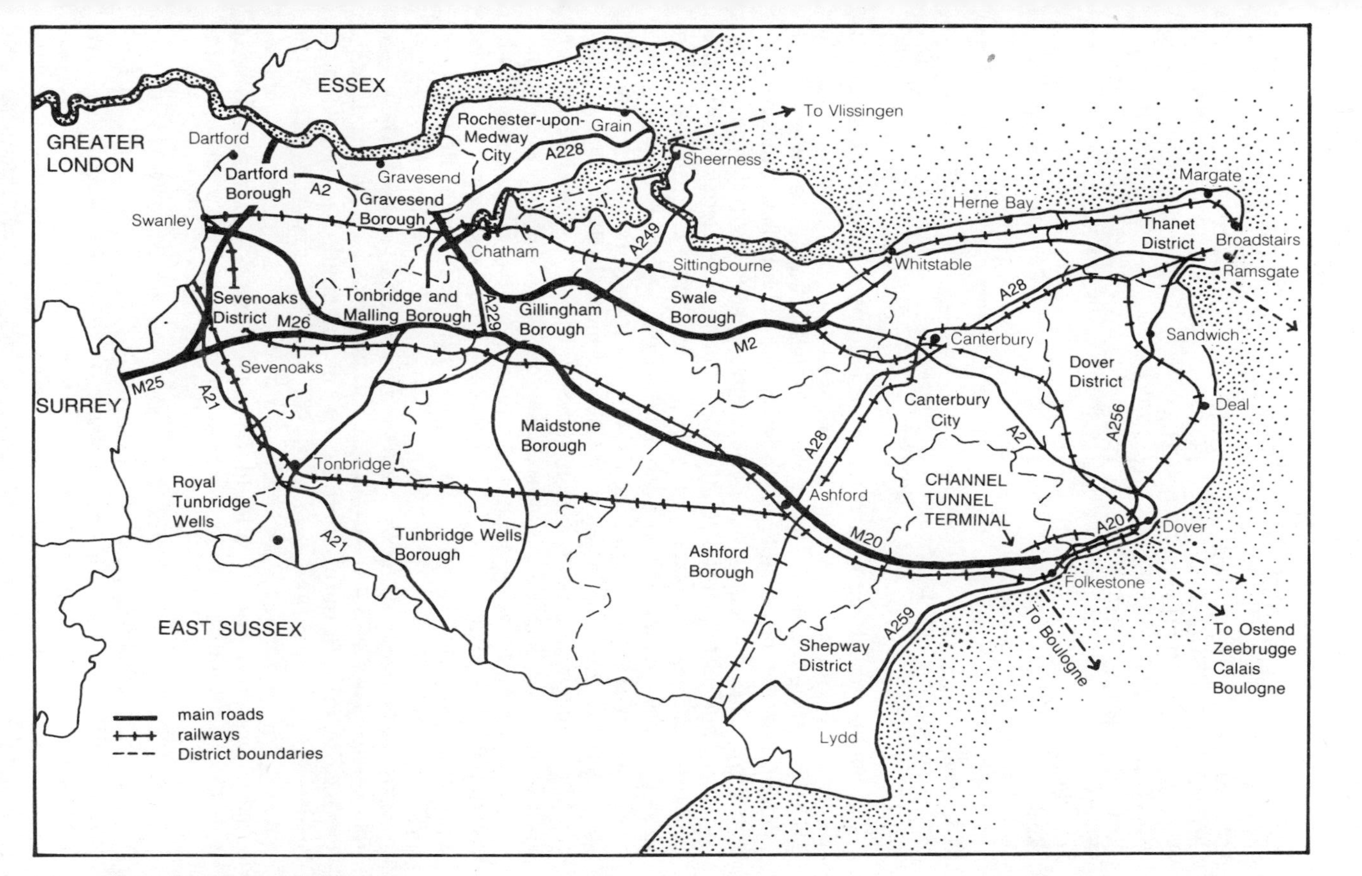

Figure 4.2 *Situation map: Kent*

Table 4.1 *Average gross weekly earnings in South-East England (£)*

	Manual		Non-manual	
	Men	Women	Men	Women
Gtr London	208.60	134.40	322.10	192.60
Berkshire	193.70	130.10	301.60	166.30
Surrey	184.10	124.60*	280.30	162.00
Bucks	192.60	125.80*	275.70	168.20
Beds	195.00	123.80	273.00	151.80
Essex	192.40	115.50	262.70	149.30
Hampshire	188.50	116.10	261.40	156.40
Kent	183.90	113.20	250.10	147.70
East Sussex	168.50	105.30*	232.70	146.20

* Estimate
Source: Kent Economic Development Board (1988).

As may be seen from Figure 4.2, Kent is located in the extreme South-East of England, and has an extensive seaboard. To the west, it borders East Sussex and Surrey. However, its most important neighbour is London, to the north and north-west, which dominates the county's economy, and in many ways effectively isolates it from the rest of Britain. Thus, although the county shares many characteristics with other parts of south-east England — being in many respects a rural and sub-urban county, in part stockbroker belt, in part village England, in part small industrial town or suburban sprawl — it operates at a significantly lower level of economic activity and development. In 1987, average gross weekly earnings were the lowest in the South-East, with the exception of East Sussex (see Table 4.1). Indeed, wages in Kent were lower even than the national norm, being on average (for full-time employment including overtime) £214.20 for men (as compared with £224.00 nationally), and £140.70 for women (£148.10) (Kent Economic Development Board, 1988, p.12).

Early work undertaken for the *Kent Impact Study* shows that in 1981 the county had a higher proportion of employees — 36 per cent — engaged in manufacturing than the rest of the South-East, and, indeed, than the rest of Britain, but little of this industry was large-scale. Instead, the county relied on small- and medium-sized businesses, and on a large service sector. In addition, 83 300 workers commuted daily to London, accounting for 13.6 per cent of employed workers in the county.

Within these global figures, the division of industry by growth and stagnant sectors is important (Vickerman, 1987a; 1987b). The East Kent economy, in particular, shows strength in nationally declining sectors, and weakness in nationally growing sectors.

In essence, the county's economy could — and can — be divided geographically into three. North Kent is its principal manufacturing area, and suffered from recession in the 1970s: the reduction in employment here between 1971 and 1981 compares unfavourably with growth of nearly 10 per cent over the same period in the rest of the South-East (outside Greater London). Mid- and South-West Kent is a more typical part of the Home Counties, with a very large proportion of London commuters (21.5 per cent), and employment growth between 1971 and 1981 of some 16 per cent. East Kent, distanced from London and historically reliant on tourism (from the capital), now has an uncertain economic base, as the tourist industry has moved into decline with the rise of foreign travel and the package tour. Although East Kent achieved employment growth of 4 per cent between 1971 and 1981, this remained below the regional average, and by the beginning of the 1980s it was clear that the area was economically one of the most depressed parts of the UK. Unemployment in the county as a whole in March 1986 was put at 12.1 per cent (compared with 10.1 per cent for the South-East, and 13.6 per cent for Great Britain). Within Kent, unemployment was 12.7 per cent in North Kent, 7.9 per cent in Mid- and South-West Kent, and 15.0 per cent in East Kent. Within East Kent, it was 12.0 per cent in Ashford, 14.2 per cent in Canterbury, 13.1 per cent in Dover, 16.1 per cent in Shepway, 13.2 per cent in Swale, and 21.4 per cent in Thanet (Channel Tunnel Joint Consultative Committee, 1986, pp.24–40).

Thus, at the start of the 1980s, when the idea of a fixed link was again mooted, parts of Kent had economic problems which in some ways matched those of Nord-Pas de Calais. Indeed, a synthetic index (GDP per capita, GDP per employee and labour-market conditions) of European regions in 1985, prepared by the European Commission, set Kent and Nord-Pas de Calais right next to each other on 97 (EC = 100) (Commission of the European Communities, 1987). However, what the apparent similarity of economic situation and the clear identity of synthetic index masked was a crucial disparity of both economic structure and industrial location. In contrast to Nord-Pas de Calais, Kent had no tradition of heavy industry, and its most depressed parts were significantly distanced from the likely locus of fixed-link benefit. Thus, whereas Nord-Pas de Calais could expect both a revitalisation of its manufacturing base in the construction phase of a fixed link, and a repositioning of its commercial heartland in the operational phase, Kent would make little apparent gain in either phase.

This statement needs to be partially qualified. At the beginning of the 1980s, the general theme of the Kent Structure Plan (approved by the Secretary of State for the Environment in March 1980 and December 1983) was growth in economic activity, and a particular theme for many years had been designation of Ashford as the county's growth point. A Channel fixed link was likely to further both of these objectives, in general by increasing economic activity in Kent, and specifically by routeing traffic to the Continent through Ashford (which was feasible with almost any proposed scheme). However, other aspects of the county's strategy were likely to be less well served by a fixed link. In particular, further pressure would be put on Green Belt areas of West Kent, and, more importantly, major job losses would be sustained in the ports of

Table 4.2 *Employment projections in Kent's ports, 1985–2003*

	1985	1993	2003
No Channel Tunnel: port employment	12450	13200–13800	14000–15400
[1] Channel Tunnel: port employment only	12450	6600–9500	7400–11300
Reduction in port employment		6600–4300	6600–4100
[2] Channel Tunnel: port and Tunnel employment	12450	9800–12700	11100–15000
Reduction in port and Tunnel employment		3400–1100	2900–400

Source: Channel Tunnel Joint Consultative Committee (1986).

East Kent, which were the mainstays of already weak local economies. Initial estimates by the Kent Impact Study Team of the impact of the Channel Tunnel on county employment confirmed fears of job losses in Kent ports, as shown in Table 4.2.

In these circumstances it would be wrong to say that reaction in Kent to a Channel fixed link was overwhelmingly either positive or negative. Certainly it was not the former: in port towns, and in rural parts of the county, the fixed link was often viewed with great hostility; and the island mentality of the British, and their traditional latent xenophobia, remained developed enough in Kent (and throughout the country) to ensure that many people without a direct interest in the project were opposed to it. Equally, however, it was not the latter: a number of important interests, including Ashford Borough Council and the town's Chamber of Commerce, supported the scheme. Beyond Kent, a fixed link was viewed as a good thing by business in general, and by the construction industry in particular, which consistently lobbied government to revive the project.

Initially, official reaction in Kent expressed these divided opinions by doing nothing. To this day, Kent County Council has not taken a position for or against the Channel Tunnel, though it would be quite inaccurate to claim that its do-nothing policy has been maintained. Indeed, since 1985 the County Council has been extremely active in attempting to maximise the benefits and minimise the costs to Kent of a fixed link. However, this still places it at least three years behind its French equivalent, the Nord-Pas de Calais Regional Council, and it leaves it in the slightly awkward position of seeing both good and bad in the Channel Tunnel.

4.3 North-west Europe

The preceding discussion has indicated that while it is just conceivable that the French agreed to build a Channel Tunnel simply in order to revitalise and reposition the regional economy of Nord-Pas de Calais, the British most certainly must have had wider ambitions. If only Kent's interests had been taken into account, the fixed link would not — perhaps ever — have been built. In part, these wider British ambitions were undoubtedly political: to be seen to be 'doing something' for Europe; and to give the private sector a visible and glamorous means of proving its superior efficiency and effectiveness to the public sector. In part, however, they were economic: to improve transport links between Britain and its EC partners, which were rapidly becoming both important export markets and key import sources; and thereby to link the national economy into the expanding economy of the EC. These wider economic ambitions are expressed in the preamble to the Tunnel treaty:

> Confident that a Channel fixed link will greatly improve communications between the United Kingdom and France and give fresh impetus to relations between the two countries,
>
> Desiring to contribute to the development of relations and of exchanges between the Member States of the European Communities and more generally between European States . . . (HMSO, 1986, p.2).

They require us to situate the Channel Tunnel in its extended hinterland, and to complement bilateral consideration of the regional economies of Nord-Pas de Calais and Kent with assessment of an extended economic space, as a basis for subsequent comparative analysis of public policy.

It is difficult to identify the precise boundaries of the Channel Tunnel's extended hinterland, but safe to say that, on the one hand, it does not reach far beyond the limits of north-west Europe (however conceived), and that, on the other, there are few parts of north-west Europe which will be entirely unaffected by construction and operation of the Tunnel. This part of Europe is currently the commercial and industrial heartland of the EC. It is dominated, and in some senses defined, by the 'golden triangle' of Greater London, Paris/Ile de France, and the Rhein–Ruhr area of Germany. Contained within the golden triangle (and its many satellites) are perhaps 100 million of the EC's (post-German unification) 340 million people, and an even greater proportion of its value-added and wealth.

Two aspects of the north-west European economy are particularly salient to discussion of the Channel Tunnel. One is that inside the golden triangle lie regions of comparative disadvantage. As can be seen from Figure 4.3, these tend to be located in peripheral regions of the national economies which the EC's 1992 programme and the Channel Tunnel will together partially undermine: in Essex, Sussex and Kent, along the Franco-Belgian border, and along the Belgian–German border. The other is that outside the golden triangle lie regions of comparative disadvantage (also shown in Figure 4.3) which are effectively isolated from the EC's main heartland by the barrier of the three major conurbations themselves. This is particularly true of regions around London and Paris, which act as screens to potential entrants to the golden triangle. This screening effect is particularly damaging to regions beyond London,

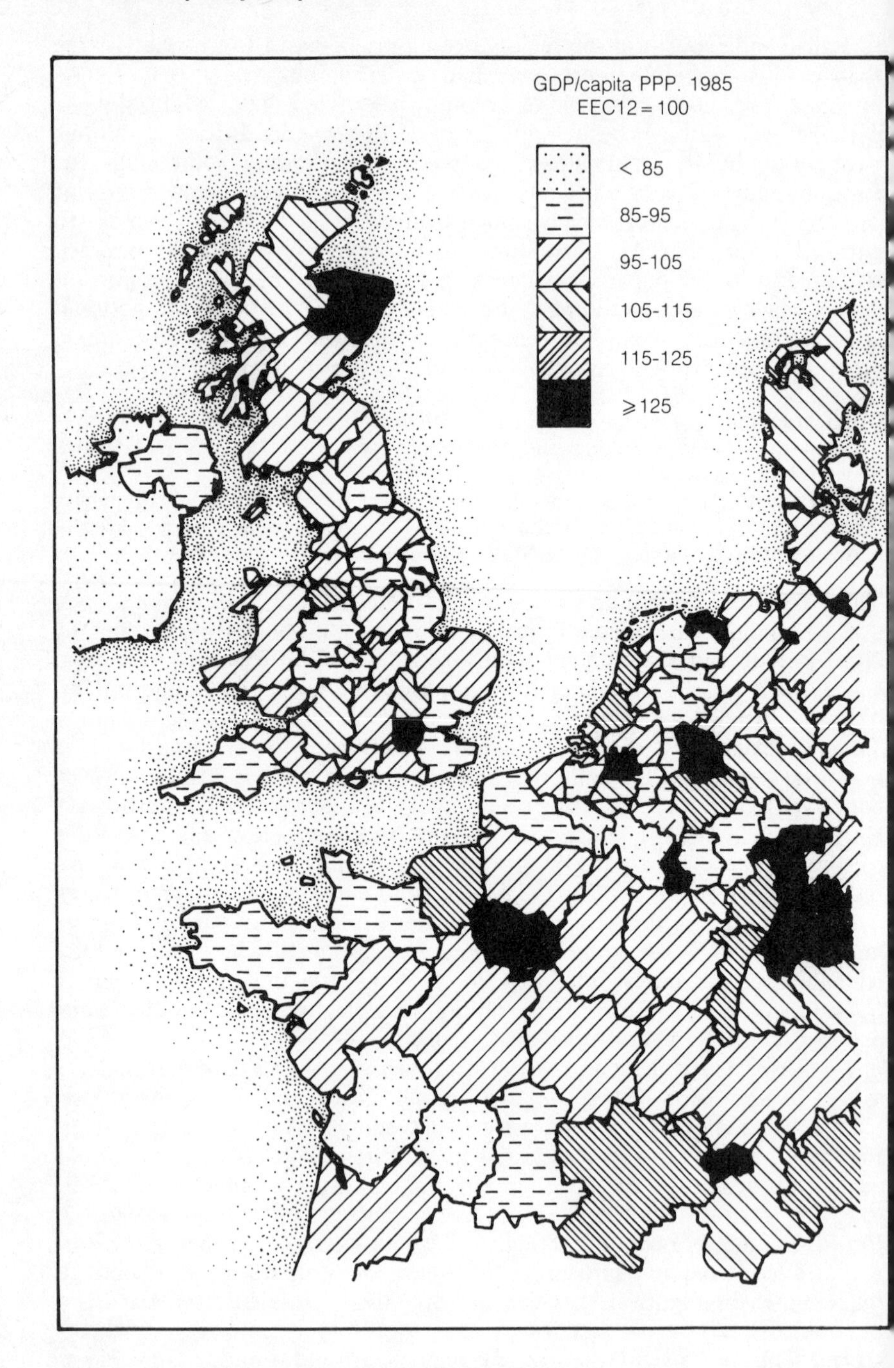

Figure 4.3 *Economic status: regions of north-west Europe*

which, unlike regions on the mainland of Europe, have few alternative natural trading partners.

The extended economics of the Channel Tunnel therefore comprises attempts by regions within the north-west European heartland both to profit from the construction activity that the Tunnel will generate, and (more importantly) to link themselves into the new transport system which will be established with its opening. In the decision phase, for example, great play was made by British and French steel-makers of the benefits construction of a cross-Channel bridge would offer their desperate industries. (Given the amount of public money routinely poured into both industries, and particularly the French, it is a measure of this scheme's impracticality that it was not selected.) Similarly, President Mitterrand is said for a while to have favoured the EuroRoute scheme in part for these reasons: that (partly because the consortium had a large French presence) it would create many jobs in depressed parts of France, not just in Nord-Pas de Calais but also in Lorraine and elsewhere.

Attempts to secure links to the new transport system being constructed around the Tunnel have been made by disadvantaged regions both inside and outside the golden triangle. Inside it, Amiens, capital of the Picardie region of France, which is both as crisis-ridden as Nord-Pas de Calais and as dominated by a national capital as Kent, has fought hard to challenge its exclusion from the TGV Nord line, just as Stratford in east London has fought — and continues to fight — for a passenger station and freight terminal on the high-speed rail link from Cheriton to London. Outside it, the specifically Tunnel-related struggle has been conducted chiefly in Britain, as regions to the north, north-west, west and south-west of London have campaigned for improved transport links to the Tunnel and, through it, to mainland Europe. In France, however, and elsewhere in Europe, campaigns to secure location on the high-speed rail network have also been mounted with some determination (CEDRE, 1990).

Each dimension of this extended hinterland needs to be understood if policy responses to the Channel Tunnel are to be properly assessed. As has been noted before, this requirement is most stringent in the UK, where policy responses simply cannot be evaluated merely on a regional (and certainly not on a county) basis.

Notes

1. Hayward (1986, pp.96–7) notes that 40 billion francs of public money were poured into Usinor and Sacilor, its Lorraine-based equivalent. In the debate on steel nationalisation in October 1981, Budget Minister Laurent Fabius was to call this the 'greatest scandal since Panama'.
2. This account draws heavily on Hayward (1986, Chapter 5).

5 *Institutions, actors and the making of public policy*

An investigation of policy responses to the Channel Tunnel in Britain and France presupposes two contexts, each of which has already been discussed in this book. The first is the policy framework in which the Tunnel scheme was itself devised. The second is the economic situation in which policy is developed. Chapters 2 and 4 cover this ground. The purpose of this chapter is to bring the two contexts together in a discussion of the institutions which have structured policy responses to the Tunnel, and of the actors who have been prime movers in policy development. This requires successive analysis of policy networks and policy communities; fixed-link policy-making in Britain and France; and British and French policy styles. On the basis of this analysis, policy outputs in the case of the present Tunnel project will be investigated in Chapters 6 and 7.

5.1 Policy networks and policy communities

Many recent studies, notably those conducted by Richardson and Jordan (1979) and Rhodes (1988) in the UK, and by Hayward (1986) in France, emphasise the central role played by policy networks and policy communities in contemporary public policy-making. In this, they often draw explicitly on the pioneering study of British policy-making undertaken by Heclo and Wildavsky (1981). The two terms 'policy network' and 'policy community' are related but distinct, the one being an extension of the other.

Thus, a policy network may be defined as any group of actors involved in an identifiable policy-making domain or process. A policy community is altogether more developed than this, being a policy network which has attained, at least temporarily, a degree of stability and permanence. The important difference is that at this point of crystallisation into a stable policy community, frontiers begin to be raised around a policy network, and the possibility of outside influences (including political change) penetrating the policy domain is correspondingly reduced. Rhodes (1986, p.23) thus writes of policy communities that they are relatively stable with continuity of membership. Decisions are taken within communities and this process is substantially closed commonly to other communities and invariably to the general public (including Parliament). Policy communities are, therefore, stable integrated policy networks.

When a policy network attains a degree of stability and permanence, then it may be identified as a policy community. At its most basic, a policy community is a semi-permanent or permanent policy network.

These two terms will be used in this chapter to structure discussion of Channel Tunnel policy-making. Their value is that they impose some sort of analytical rigour on policy-making which has often been the result of developed contacts between government bureaucrats, on the one hand, and interest groups, on the other. However, as will be shown here, the nature and extent of these contacts has been very different in Britain and France.

Given the project's long history, it is no surprise to discover that the policy network with an interest in a Channel fixed link has altered — and expanded — over time. In the nineteenth century, in *laissez-faire* Britain, and even in *colbertiste* France, policy was determined first, tentatively, by railway barons seeking new markets for their services, and second, definitively, by the British military establishment seeking preservation of Britain's island status. In the twentieth century, as the British defence community has relinquished its control, fixed-link policy has similarly been given an important impetus by construction companies in both Britain and France, but it has also been driven to varying extents by more diverse groupings, ranging from regional industrial interests in different parts of Britain and France, to national commercial interests in the two countries, and even to supranational political interests. In this latter context, the decision of DG VII, the Transport Directorate-General of the European Commission, to sponsor a study of fixed-link options in the late 1970s (Commission of the European Communities, 1980) may be cited: it was a clear attempt to launch a Community transport infrastructure programme by means of a high-profile, and evidently *communautaire* policy. This initiative was further pursued in the 1980s (House of Lords, 1989).

However, although regional, national and supranational policy communities have sought to influence fixed-link policy in both Britain and France, the asymmetry in economic situation between the two countries, detailed in Chapter 4, has generated an equivalent asymmetry in the fixed link's political situation. In Britain, the policy community with a positive interest in fixed-link development has consistently been almost exclusively national. In France, throughout the period since 1955, when the screen of the British defence argument was lifted, it has consistently had an important regional, as well as national, dimension.

It is important to note that this asymmetry is in some ways a structural feature of the two national political and administrative systems: it would have been observed to some degree whatever the regional economic situation of the fixed link in Britain and France. In France, the *cumul des mandats*, already referred to, and illustrated most spectacularly by the case of Mauroy, combined with extensive administrative links, ensures that localities are structurally represented at the centre of French government in a way that is not replicated in the UK. The most developed presentation of this difference between the two systems is Ashford's (1982) British dogmatism and French pragmatism thesis, which reverses traditional judgments of the conduct of policy in the two countries, and argues that the French system, by being vertically integrated, is flexible and effective, and that the British, by being divided on strict horizontal lines, has neither of these qualities.

Based as it is on work undertaken in the 1970s, Ashford's thesis does not take account of the decentralisation reforms introduced by the incoming Socialist government, under Prime Minister Mauroy, in the years after 1981. Centred on the *loi Defferre* of 2 March 1982, these reforms substantially altered the legal status and responsibilities of French governmental institutions, but had less impact on their actual operations, and on relations between their constituent parts. With the emergence in the postwar period of strong towns and administrative regions, institutional practice had already begun to diverge from institutional texts, and to demonstrate a marked decentralist bias. As Mény (1987, p.57) writes:

> In some respects, the law was a simple codification of the practices of many left municipalities, *départements* and regions before 1981. The most striking changes — suppression of *tutelle* and the transformation of the prefects' function — have only been partially implemented.

As a leading bastion of the Left, Nord-Pas de Calais had been at the forefront of *de facto* decentralisation in the years prior to 1982.

Similarly, Ashford's thesis pre-dates the major programme of institutional reform conducted by successive Thatcher governments in the years 1979–90. However, few would argue that this programme was designed to undermine the marked centralist bias of British political institutions. Indeed, it can be argued that British local government legislation in the 1980s actually confirmed the existing reality of central–local relations. By this is meant that British local government has always been vulnerable to the sorts of change fashioned by central government in the 1980s, and that economic crisis of the kind witnessed in the mid-1970s was always likely to provoke them (Bulpitt, 1989). The uniqueness of Mrs Thatcher was simply her refusal to shelter behind the old formulae of central–local 'partnership', which never contained much truth.

Hence, in reconsidering his thesis at the end of the 1980s, Ashford (1989) was able to argue that, if anything, it was more strongly grounded in the politics of the late 1980s than it was in those of the late 1970s. French government remains flexible. In Britain, by contrast, despite deep politicisation of local government, and of central–local relations, 'virtually impenetrable barriers between national and local government' (Ashford, 1989, p.78) continue to exist.

This thesis provides important insights into the politics of the Channel fixed link. Considering only the 1980s scheme, it is evident that the French policy community was far more extensive than was the British, having a developed regional dimension in Nord-Pas de Calais which simply had no British equivalent. The Nord-Pas de Calais policy community was not simply focused on Mauroy, though he was obviously an important element in it, but extended, partly through *notables*, and partly through the administrative services of the state, into all parts of the region. Detailed differences between Britain and France are best illustrated through investigation of fixed-link policy-making in the two countries.

5.2 Fixed-link policy-making in Britain and France

A discussion of prime movers in launching the present fixed-link scheme must, in fairness, include the various entrepreneurial and commercial interests which, since the 1870s, have promoted the idea of a fixed link, and ensured that in the end the British government would have to abandon its increasingly implausible defence objections and allow realistic consideration of fixed-link options to proceed. Indeed, throughout the period of militaristic veto on the British side, from the 1880s to the 1950s, a single private-sector interest, the Channel Tunnel Company, formed by Sir Edward Watkin, a frustrated tunneller in the 1880s, remained in existence to keep alive the idea of a Channel Tunnel. With the defence argument undermined, this company was to partner government in the 1970s scheme. Thus, in a sense, the fixed-link project has had a consistent commercial drive, which has ensured its periodic appearance on the policy agenda in both Britain and France.

However, looking in more detail at the present scheme, it is clear that the project was relaunched by more specific factors: an act of attempted Euro-entrepreneurship on the part of DG VII in the late 1970s; and the sound commercial instincts of BR and SNCF in following this up with a viable scheme submitted to government in 1979. At this point, the British government became interested, and officials in the Department of Transport entered the policy network, only to increase its extent considerably by (re)opening the whole issue to private-sector interests. The French government, still suspicious of British intentions, kept its own counsel.

Thus, commercial and industrial interests entered the policy network only on the British side at the beginning of the 1980s. Indeed, such was the extent of French *méfiance* that it would be hard to identify a French element in the network at this stage. This situation was to be changed by the Thatcher–Mitterrand announcement of an Anglo-French Study Group in September 1981. At this point, the French fixed-link network rapidly expanded to its full national and regional extent. Guy Braibant, the only Communist *conseiller d'Etat*, was asked by Communist Transport Minister, Charles Fiterman, not only to head the French side of the Study Group but also to conduct a fact-finding mission in Nord-

Pas de Calais. In November 1981, this mission conducted soundings successively in Calais (which is Communist-controlled), Boulogne, Dunkerque and Lille, in each town meeting political leaders as well as representatives of industrial and social interests brought together through the various semi-public *chambres* and *syndicats* which exist in France, and through the Comité Economique et Social Régional (CESR).

Simultaneously, two Working Groups were established. One, instituted by the Regional Council, under Senator and Regional Council vice president, Daniel Percheron, was to produce a 'political' report setting out regional requirements of a fixed link. It worked in close collaboration with the CESR. The other, established by the prefecture on the initiative of the Transport Ministry, was to reproduce the work of the Braibant Commission at a regional level by taking soundings throughout the region (and not just in its four main towns). It brought together the regional chamber of commerce, and local and national officials. Importantly, both groups worked in close collaboration with groups outside the Regional Council, and with staff at the regional prefecture in Lille (which even collaborated in preparation of the 'political' Percheron Report). Indeed, the prefecture became the main central–local co-ordinating mechanism, working both with key officials in the Ministry of Transport and with officials in the Regional Council. In this way, extremely close *concertation* between centre and periphery was generated in early French discussions of the present fixed-link project.

In Britain at this time, the regional dimension to Department of Transport (DTp) deliberations was almost entirely non-existent. The project was mentioned in the course of normal contacts, and officers from Kent County Council (KCC) sometimes sought to force civil servants to consult with them through the various pressure points — some proper, some improper — that are known to experienced local government officers. However, there was certainly no developed attempt to involve Kent in official consideration of fixed-link options, and the off-the-record personal briefings which civil servants from time to time gave selected Kent officers in no sense compare with the vast network of official contacts which existed in Nord-Pas de Calais at this time. Indeed, the DTp, acting within tight political constraints, was determined to exclude local government from its discussions as much as possible. Not until the time came to prepare the *Invitation to Promoters*, in late 1984 and early 1985, did DTp officials establish formal contacts with KCC and the implicated districts, and then it was largely to inform them of decisions taken rather than to involve them in deliberations.

The difference between British and French fixed-link policy networks is perhaps most graphically illustrated by a meeting which took place in Calais in the period between 2 April (*Invitation to Promoters*) and 31 October (deadline for submissions) 1985. This was between senior planning officers from KCC, and senior officials from the French Ministry of Transport, who, unable to get what they considered necessary information on the British periphery from British DTp officials, sought to establish direct contact with Kent. From this time onwards, Kent officials were part of the French fixed-link network, but continued to find it difficult to penetrate the British network.

This disparity was undermined to some extent after the *Invitation* had been issued in April 1985, as, in both Britain and France, tendering consortia were required by government to undertake consultation exercises at the local level in order to demonstrate the local acceptability of their rival schemes. These discussions were a great deal more extensive in Kent than in Nord-Pas de Calais, for the simple reason that the decision in principle (in favour of a fixed link) had already been secured in Nord-Pas de Calais, whereas it had not been seriously considered in Kent.

However, even at this stage, when even the British periphery had been brought into fixed-link discussions, the major difference between Britain and France remained that Kent was required to react to a policy decision which in France had been taken in close consultation with Nord-Pas de Calais. Behind the scenes, substantial disparities remained apparent on the two sides of the Channel. In France, state officials continued to collaborate with officials from the Regional Council in preparing a set of regional requirements of a fixed link, such that when the resultant dossier was presented to President Mitterrand on 20 January 1986 (Conseil Régional Nord-Pas de Calais, 1986a), the Regional Council knew that some 75 per cent of its demands would be met by government. The Council of Ministers, meeting two days later, duly confirmed this proportion of acceptable measures. No parallel collaborative activity was undertaken in Britain.

Analysis of fixed-link policy networks in Britain and France thus confirms Ashford's thesis that the periphery is structurally represented at the centre of French government in a way which is not approached by the British system. Indeed, on the British side fixed-link policy-making forcefully recalls Heclo and Wildavsky's investigation of the expenditure process in central government, dominated by 'family life' in the Treasury, and by 'village life' in Whitehall, which asserts that in Britain 'governing public money is a private affair' (Heclo and Wildavsky, 1981, p.2).

By this Heclo and Wildavsky mean that large tracts of British public policy are determined by a private society of London-based civil servants, presided over by the Treasury, which is effectively closed to outsiders. Within this society, face-to-face (or telephone) contacts predominate. In the case of the Channel Tunnel, political will was a crucial element in British policy-making. Otherwise, the picture drawn by Heclo and Wildavsky — and particularly its description of village life in Whitehall — is an accurate portrayal of fixed-link policy-making in Britain. It was, indeed, a closed world, open on occasion to potential fixed-link builders and operators, but only under duress to the local authorities which would be implicated in a decision to build.

The clear strength of this system, mentioned before in this book, and identified by Heclo and Wildavsky, is the high degree of coordination at its (very) centre, much of it provided in a quiet but efficient manner by the Cabinet Office (Seldon, 1990). However, this central capacity is bought at the price of coherent decision-making and implementation throughout the government system. 'The strength of British political administration . . . [is] in the solidarity of its upper echelons; its weakness [is] in the sacrifice of substantive rationality (policy results) to maintaining its communal culture' (Heclo and Wildavsky, 1981, p.1). This subordination of policy to community is all the more grave when tiers of government are necessarily implicated in policy implementation, yet systematically excluded from policy-making.

Very different institutional attributes are found in France, though here, too, a succession of authors has described the close personal relations which exist at the centre of the state. Thus Suleiman (1974; 1978) maps the contours of close-knit French elites, and shows them to be in clear control of the main organs of state power. Similarly, writers from the Crozier school have detailed many aspects of *The Bureaucratic Phenomenon* (Crozier, 1964), and in a more sinister vein, critics of the regime have attacked *L'Ere des technocrates* (Thoenig, 1973), *The Heights of Power* (Birnbaum, 1982) and *L'Enarchie* (Mandrin, 1967). The difference between Britain and France would, then, seem to be twofold. First, in France 'village life' is not restricted merely to high-ranking state officials, but incorporates elites in both the (subnational) public and private sectors through the prefectoral system, and the practice of *pantouflage* by which senior government officials move into important positions in the private sector. Second, in France the centre simply cannot ignore the periphery because in many cases the same people occupy key positions in both spheres of government.

In these circumstances, the workings of the French governmental machine can be painfully slow (particularly to a British observer).[1] Furthermore, as was mentioned above, parallel working groups were organised in Nord-Pas de Calais with direct input from the highest levels of the state services. The result is that governmental co-ordination and operation in France is nothing like as easy as it is in Britain, but there would seem to be a better chance of generating genuinely effective policy-making and implementation in France as a consequence of patient initial attention to construction of the necessary coalitions.

Before moving on, it is worth trying to determine which of the two differences between Britain and France specified above is the primary factor in necessarily extensive French coalition-building. While great stress is often placed on personal links between the public and private sectors in France, it is not clear that this is in fact the determining factor. Clearly, the French system of *grandes écoles* and *grands corps* is singularly elitist, and singularly effective in channelling ability into both public service and private enterprise. However, what is not clear is whether this system is either as generative of 'shared interests', or as different from the British, as is sometimes thought. On the one hand, it is by no means proven that an individual will always place the interests of a corps above that of a present employer, though both Thoenig and Suleiman have detailed cases in which this has indeed happened (discussed by Stevens, 1980). On the other, it is always possible to create extensive elite networks in the absence of common backgrounds and experience.

This last point is illustrated by the present Channel Tunnel project. In France, leading figures at Eurotunnel and TML are, as one might expect, products of the *grandes écoles*. André Bénard, co-chairman of Eurotunnel since February 1987 (and sole chairman since February 1990), is a provincial (Nantes, Marseille) who was educated at the Ecole Polytechnique in Paris before embarking on a 37-year career in the oil industry (*The European*, 1990). Philippe Essig, from the 8th *arrondissement* in Paris, and currently chairman of TML, is a quintessential product of the French technocratic elite. He was educated at the Ecole Nationale d'Administration before moving successively through a French colonial railway company, DATAR, the Paris metro system (RATP) and SNCF, of which he was made chairman in 1985. After a brief spell as Housing Minister in the first Rocard government of 1988, he was appointed chairman of TML in May 1989 (*The Independent*, 1990). These are impressive profiles which illustrate — and in the case of Essig almost caricature — the easy movement of, and formidable contact between, elites in the French system.

However, a similar profile may be found on the British side of the project. Alastair Morton, also co-chairman of Eurotunnel from February 1987 to February 1990 (when he was made chief executive and deputy chairman), is a South African who was educated at Oxford and MIT before moving through a number of financial institutions in London and Washington. In 1967 he joined Harold Wilson's Industrial Reorganisation Corporation, and since then has moved between the public and private sectors in ways which would be considered usual in France: 1970–6, executive director or chairman of various investment groups and public engineering bodies; 1976–80, chairman of British National Oil Corporation; 1979–82, British Steel; 1982–87, chairman of Guinness Peat; 1987–90, British co-chairman, subsequently chief executive, of Eurotunnel (*The Times*, 1987). Perhaps of most interest is the manner in which Morton has obtained his recent posts: both the Guinness Peat and the Eurotunnel appointments were arranged by leading City financiers, such as Sir Kit McMahon, former Treasury official, now chairman of Midland Bank, and David Walker, at the Securities and Investment Board. In each case, the discreet presence of the Bank of England at the centre of operations was important to success (*Spectator*, 1987b; *The Independent*, 1989b).

Indeed, by spring 1990 the Bank of England had intervened at least five times to help Eurotunnel overcome difficulties. First, in the early 1980s and again in autumn 1986, it leant on banks to support the project, and, in the latter case, to ensure successful completion of Equity II. Second, in October 1986, it was instrumental in the appointment of Sir Nigel Broackes, chairman of Trafalgar House (the lead shareholder in EuroRoute), as a non-executive director of Eurotunnel in an attempt to inject a new management style into a struggling project. Third, on the resignation of British co-chairman, Lord Pennock, in February 1987 (followed a week later by Broackes's resignation), it helped to resolve a management crisis by head-hunting Morton (having, it is said, previously approached Sir John Harvey-Jones of ICI and Sir Jeffrey Sterling of P&O (*Spectator*, 1987a)). Fourth, in January 1990, it is believed to have called a meeting with Eurotunnel's bankers to resolve a costs and consequent financing crisis. Fifth, in February 1990, Robin Leigh-Pemberton, former chairman (until 1983) of National Westminster Bank (a founder shareholder in CTG) and at the time governor of the Bank of England, chaired a meeting between Eurotunnel and TML designed to overcome an increasingly bitter dispute over costs, delays, and management (*Financial Times*, 1990a). It was this meeting which resulted in Morton's move from co-chairman to chief executive of Eurotunnel, and in a number of new appointments (some of them American) to top positions in the British management hierarchy. Prominent among the new appointees were John Neerhout, executive vice president of the US construction group Bechtel, as project chief executive; Alistair Fleming, former BP Exploration projects manager, as managing director for construction; and Keith Bernard, a senior executive of Bechtel, and formerly with Bay Area Rapid Transit (BART) in San Francisco, as managing director for development of the tunnel's transport system (*Financial Times*, 1990b).

Morton's experience, and this series of Bank of England initiatives, suggest that even though Britain has neither *grandes écoles* nor *grands corps*, public–private elite networks are certainly developed in the UK, if not quite to the same degree as their French equivalents. As was noted above, the Whitehall village at least communicates with leading parts of the private sector, and would seem on occasion to incorporate key elite members of it. Once the further point is made that public and private elites in both Britain and France may well be equally prone to identify more with their present employer than with their past experience, then the difference between the British and French systems in this regard would not seem to be substantial. In brief illustration of this last point, in 1987 Essig, as chairman of SNCF, won state finance from the free-marketeer Chirac government both to upgrade the national rail system and to finance (though not subsidise) TGV Nord, when many believed the line should be privately built and financed (*The Independent*, 1990). Similarly, Machin and Wright (1985, p.21), in a review of the early years of the Mitterrand presidency, list a number of conflicts between state and industry: 'the semi-autonomous baronies of the SNCF, EDF, Elf-Aquitaine — the "States within the State" — proved no more amenable to left-wing control than they had to that of previous régimes'. This does not suggest that agreement between the French state and its industrial placemen is always total.

The main lesson of this comparison of state–industry links surrounding the present Channel Tunnel project is, then, simply that modern governments like to spin complex webs in order to exercise at least a modicum of control over schemes of the magnitude and strategic importance of the Tunnel. It is hard to believe that great differences between the British and French political systems are visible here.

Instead, the crucial institutional difference between Britain and France would seem to be the structural representation of the French periphery at the centre of government, enshrined in the dual *cumul des mandats* and prefectoral systems, and enhanced by the system of ministerial *cabinets*, established under the Restoration, and a key aspect of contemporary French government. These *cabinets* now often contain regional representation, and give the periphery direct access to the centre of French government. Certainly, Mauroy's prime ministerial *cabinet* in 1981–4 contained members whose primary territorial loyalty was to Nord-Pas de Calais. Thus, the constitution of the Fifth Republic, which compels Deputies and Senators to resign their parliamentary seat when they are made ministers, does not appear to have undermined the territorial dimension to French national politics (de Lamothe, 1965).

It is the existence of these various mechanisms in France, and their almost complete absence in the UK, which ensures that policy networks stretch from centre to periphery in France, and that, if need be, the periphery can be wholly ignored by the centre in the UK. This structural difference has already been discussed. All that needs to be stressed here is that localities are guaranteed access to national policy-makers in France, in contrast to the situation in Britain, where no such guarantees exist. The value of French access, and the penalties imposed by its lack in Britain, will be reviewed in chapter 8.

5.3 British and French policy styles

On the basis of observed differences in policy networks and communities in Britain and France, a number of commentators have attempted to construct a typology of policy styles in the two countries. In a series of articles, Hayward (1974; 1976; 1982) contrasts two styles of policy making, 'humdrum' (which draws on Lindblom's (1959) science of muddling through) and 'heroic', which he says are respectively characteristic of Britain and France (Hayward, 1974).

Hayward's argument is not that British policy-making is always humdrum, and that French is always heroic, but that these are the forms of decision-making to which policy-makers in the two countries usually aspire. The British, deeply committed to liberal values, and possessing a strongly departmental government system, tend to favour incrementalism and mutual adjustment between competing interests, each of which possesses considerable veto power. As Richardson *et al.* (1982) note, the British style demonstrates a marked predilection for consultation, and a strong desire to avoid challenging entrenched interests. By contrast, the French, operating in a tradition of decisive state economic intervention which stretches back to pre-Revolutionary times, and driven by an informal central nucleus of real executive power-holders, tend to favour both rational planning and naked assertion of political will on the part of central policy-makers. Though marked by immobilism and stagnation for large parts of its history — particularly, as Hoffmann (1974) notes,

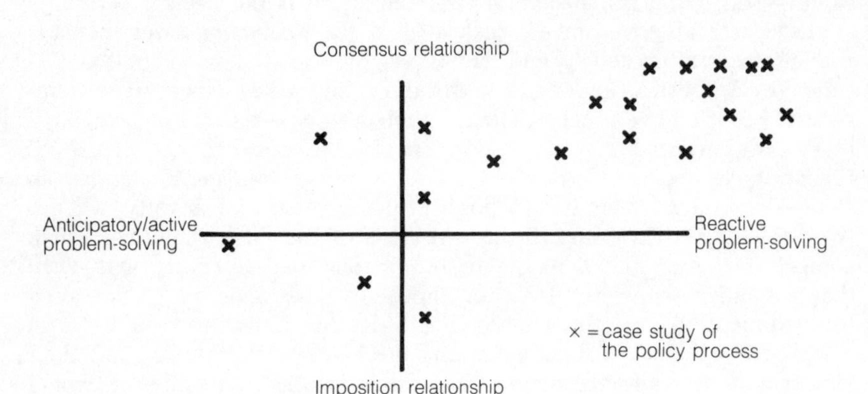

Figure 5.1 *Policy systems and styles*
Source: Richardson (1982)

during the Third Republic — French public policy is also capable of assertive action in pursuit of radical change (Richardson *et al.*, 1982).

Hayward (1982, p.116) sums up his assessment of French policy-making by stating that

> even if this assertive, active policy style does not in practice mean integrated or decisive action, it implies a *capacity* for policy initiative, a *potential* for far-sighted planning and a *propensity* to impose its will when this is necessary to obtain public objectives.

Although examples of contemporary French immobilism and policy mess may be cited — Hayward (1982) himself cites the failed Rhine–Rhône canal project — holders of central state power have the ability to engage in heroic policy-making, and the capacity to realise their chosen projects.

Applied to the Fifth Republic, this thesis has been given positive and negative interpretations. Indeed, in the mid-1960s both variants were on offer, as British commentators — best represented by Shonfield (1965) — celebrated the apparent ability of the *dirigiste* French state to direct the national economy through an indicative planning system, in stark contrast to the clear inability of the British government to do likewise, and as French commentators — led by Crozier (1964) — denounced the rigidity and authoritarianism of a political system run by an unholy alliance of bureaucrats and political *notables* in whatever they judged to be the national interest. At this time successive British governments themselves attempted to imitate French practice, but, as Hayward (1974, p.405) notes, British flirtation with French methods, born of financial crisis in July 1961, and killed by it in July 1966, was never more than 'toothless tripartism'.

Applied to an area of policy with clear local or regional implications — such as, in France, the present Channel Tunnel project — the thesis requires clear qualification, for what has been shown here is that one of the important differences between British and French policy-making is that the French periphery, in contrast to the British, has guaranteed access to the centre of state power. This suggests that national policy-making needs to be judged along two dimensions, the incrementalist–rationalist, and the closed–open. Richardson *et al.* (1982) use these two dimensions — they label the second consensus–imposition — to construct a grid (shown in Figure 5.1) by means of which policy systems may be mapped, and their predominant style assessed.

It should be noted that this grid implies an important modification of Hayward's thesis. The closed–open dimension, if salient across a range of policy issues (as it is in France, depending on whether or not policy has a marked regional dimension), suggests that there are some issues which cannot be treated heroically if opposed by significant peripheral interests. It also suggests that even humdrum policy-making, premissed on consensus and accommodation, will encounter potential disruption if undertaken by a closed political system (like the British with respect to local government). These points are developed further in Chapter 8.

Returning to the distinction between policy networks and policy communities developed at the start of this chapter, it is clear that closed and relatively stable fixed-link policy communities were developed in both Britain and France in the early 1980s. The difference between them was that in Britain frontiers around the fixed-link policy community were intended to exclude local and regional interests as much as possible, whereas in France such interests were brought into the fixed-link policy community (because they could not be excluded from it).

5.4 Conclusion

This chapter has attempted to provide both an initial assessment of the main lines of British and French fixed-link policy, and a context within which the details of policy at the local level in both countries may be evaluated. It has shown that from the start of serious negotiation of the present Channel Tunnel project, the French policy community and policy networks were a great deal more extensive than the British, having, crucially, a regional dimension which was almost entirely lacking in the UK. On this basis, it has suggested that British and French policy styles need to be placed on a two-by-two matrix if they are to be properly assessed.

Having, therefore, concluded investigation of 'high' fixed-link politics in Britain and France, and provided the necessary context for analysis of policy at the local level, we move in Chapters 6 and 7 to consider 'low' fixed-link politics in the two countries.

Note

1. See above, p.45.

6 The politics of Channel Tunnel construction in Kent and Nord-Pas de Calais

The politics of Channel Tunnel construction is heavily influenced by the economic and political contexts described in Chapters 4 and 5. Extensive French policy networks, built on a solid foundation of agreed economic imperatives and strong institutional links, permitted rapid development of policy responses to the Channel Tunnel in its construction phase. Limited British policy networks, possessing neither of these attributes, took correspondingly longer to formulate policy.

This is the first difference between British and French public policy in the construction phase of the present Channel Tunnel project. The second is that the concerns of policy-makers in the two countries were highly divergent. As a fixed link was seen as being in the interests of almost the whole of Nord-Pas de Calais, policy focused on the maximisation and division of economic gain. In Kent, far more attention was devoted to control of the Eurotunnel scheme, and to minimisation of its local costs, both economic and social (or environmental). As in many other areas of fixed-link policy, this difference may be ascribed in part to economic, and in part to institutional factors.

A third difference between British and French fixed-link policy in the construction phase of the present project may be found in the locus of policy coverage. In France, policy-makers sought chiefly to concentrate economic benefit in the Channel Tunnel's immediate hinterland, defined not simply as Nord-Pas de Calais, but within this region as Lille and all points west, particularly the coastal strip running from Dunkerque to Boulogne, and centred on Calais. In Britain, to the extent that policy-makers had a strategy to extract and direct economic benefit from the Tunnel, it was concerned above all to disperse benefit throughout the national economy.

In this chapter, each of these differences will be considered in turn.

6.1 Speed of response

Policy-makers in France were not simply quick to respond to the choice of CTG-FM on 20 January 1986, and to develop a policy to deal with a now realistic project, they pre-empted the decision by several months. As was noted in Chapter 5, the dossier of regional demands which was submitted to President Mitterrand in Lille, *Eléments pour un plan de développement de la région Nord-Pas de Calais* (Conseil Régional Nord-Pas de Calais, 1986a), contained a series of measures which had already been negotiated by state and regional officials, and which, for the most part, simply required rubber-stamping by the Council of Ministers.

The demands which the Regional Council fed into this dossier were founded on analysis undertaken for it by Bechtel France. The Bechtel report, *Impacts et perspectives pour la région Nord-Pas de Calais du lien fixe transmanche* (Conseil Régional Nord-Pas de Calais, 1985), was commissioned in February, and delivered in September and (with revisions) November 1985. The communiqué issued by the Council of Ministers consequently outlined five main areas in which 'mesures d'accompagnement de la liaison fixe transmanche' had been agreed (Conseil Régional Nord-Pas de Calais, 1986b):

(1) At an institutional level, it proposed to put into operation the *procédure 'grands chantiers'* (PGC), hitherto only activated to manage public-sector projects, in order to ensure the smooth insertion in the regional economy of such a large construction project. The main concerns of this procedure would be fourfold:
  (a) to manage problems created by the Tunnel site itself, in terms of housing, transport, education, health, and so on;
  (b) to co-ordinate local training initiatives for TML workers in the construction phase;
  (c) to inform regional firms which might be able to benefit from TML subcontracting;
  (d) to manage 'après-chantier' problems, by retraining Tunnel workers and attending to local unemployment after 1993.[1]

(2) To upgrade the road network serving the Tunnel, it proposed to ensure:
  (a) construction of a *rocade littorale* stretching from the Belgian frontier to Boulogne-sur-mer in time for the Tunnel's opening in 1993 (2750 million francs);
  (b) completion of the A26 toll motorway linking Calais to Reims by 1990;
  (c) construction of key road links to the three main Nord-Pas de Calais ports (A26 to Boulogne: 250 million francs; eastern access at Calais: 33 million francs; western access at Dunkerque: 36 million francs);
  (d) fast realisation of existing projects (notably the RN42 and A25) through additional state spending of 100 million francs per year from 1986.[2]

(3) To upgrade the rail network, it proposed:
  (a) to extend TGV Nord from Lille (which already featured in plans to link Paris, Brussels, Cologne and Amsterdam by high-speed rail) to the Tunnel mouth;
  (b) to electrify other parts of the regional network — notably the section between Hazebrouck and Calais — in order to link it into the Tunnel system (and the French national system).

(4) To modernise (and compensate) the region's ports, it proposed:
  (a) to extend the container port at Dunkerque by 1993 at the latest, at a projected cost of 200 million francs;
  (b) to extend the port at Calais, at a projected cost of 400 million francs;
  (c) to re-equip the port and the fishing fleet at Boulogne, at a projected cost of 200 million francs.

(5) To restructure the regional economy, and particularly its tourism sector, it proposed to establish a *fonds de développement du littoral*, though without specifying the level of funding.

This communiqué in turn formed the basis for a *protocole d'accord*, signed by the regional prefect and the Nord-Pas de Calais Regional Council on 14 March 1986, two days before the legislative elections which were to result in defeat for the Socialist government of Laurent Fabius and its replacement by a right-wing government under Jacques Chirac. Known as the Plan Transmanche (République Française/Région Nord-Pas de Calais, 1986), this accord set out a joint programme, financed by the state and the region, to tackle transport, training and research, tourism, and the quality of local life. Funding, when specified, was split between the state and the region on a 2:1 basis, rather than on the usual principle of equal shares.

It would be wrong to assume that each commitment written into the Plan Transmanche has since been honoured by the French state. As has been found in other regions, *contrats de plan Etat-région* have no legal standing, and can be rewritten by the state as and when necessary.[3] However, in the case of the Plan Transmanche, the only substantial departure from the initial agreement has been the decision of the Chirac government not to fund the *fonds de développement du littoral*, a decision which was reversed by the Socialists on their return to office in 1988, but which remains only a vague commitment, and has no specific finance attached to it. With this single important exception, the Plan Transmanche has been substantially realised, and public policy in France in the period since January 1986 has been chiefly concerned not with decision-making but with implementation.

In the UK, by contrast, policy-making was a major preoccupation in the period immediately after selection of CTG-FM. Unlike the situation in France, where the main lines of policy had already been decided, British Channel Tunnel policy at this time was shrouded in uncertainty. Department of Transport lawyers had started to draft what was to become the Channel Tunnel Bill, and in the event this was printed fairly rapidly, on 17 April 1986 (still more than a month after the signing of the Plan Transmanche). However, many parts of the bill — such as the central planning regime, and detailed provisions regarding access to the terminal site at Cheriton, the transport of aggregates to the construction site at Shakespeare Cliff, the disposal of Tunnel spoil, and so on — were no more than pro forma clauses which no one pretended would be retained in the ensuing Act. In each case, the issue had proved too contentious for swift resolution, and the decision appears to have been taken to resolve matters in the course of Select Committee proceedings.

Moreover, the Channel Tunnel Bill covered but a small part of the ground mapped by French Channel Tunnel policy. Partly to consider wider issues of economic cost and benefit in Kent, a Joint Consultative Committee (JCC) of civil servants, local government officers, and key private-sector interests (Eurotunnel, TML and BR) was formally established by the February 1986 White Paper (Department of Transport, 1986), and first met at the end of that same month. Its creation was not wholly original. On the one hand, the closure of Chatham Naval Dockyards (in North Kent) in the early 1980s had been handled, in part, by a similar committee (without, however, private-sector involvement), on which officials from both the Department of Transport (DTp) and KCC who were now involved in the Channel Tunnel project had served. On the other, an informal precursor to the JCC had met for a number of months before formal institution of the Committee, initially bringing together planning officers from Kent County Council (KCC) and the three directly affected districts, and representatives of CTG (and, on occasion, BR). However, formal institution was an important development because, with a Minister in the chair (the condition on which KCC agreed to participate), commitments made in committee became harder to soften, and participation became harder to dilute (through, say, the delegation of successively junior officers). The Mitchell Committee, as it became known (after its first chairman, David (later Sir David) Mitchell, then Minister of State for Transport), and its shadow officer committee, known as the Noulton Committee (after its chairman, John Noulton, then Under-Secretary at the Department of Transport), became central institutional devices on the British side of the Channel in the months after selection of CTG-FM.

A major function of the JCC was to sponsor the work of the Kent Impact Study Team (KIST), which produced, first, a *Preliminary Assessment* (Channel Tunnel Joint Consultative Committee, 1986), which was in essence a background factual report; second, a *Consultation Document* (Channel Tunnel Joint Consultative Committee, 1987a) on issues; and third, an *Overall Assessment* (Channel Tunnel Joint Consultative Committee, 1987b) based on consultants' work. Furthermore, KIST continues to produce an annual *Monitoring Report* (Channel Tunnel Joint Consultative Committee, 1989; 1990). The main KIST report was strongly criticised on publication for its alleged 1960s style and view of regional policy. However, whereas the Bechtel Report was published in full, the complete consultants' report written for KIST was never released.

Furthermore, the *Kent Impact Study*, though in many ways a political document (based on expert analysis), committed none of its sponsoring institutions to any of its recommendations. Indeed, the JCC, as its name suggests, was merely a consultative forum, with no executive power of its own. As a result, the *Study* made very few policy recommendations, and the single significant recommendation that it did make — that an East Kent Development Agency be established to cover the districts of Ashford, Canterbury, Dover, Shepway and Thanet — was undermined by the refusal, first of Shepway, and subsequently of Ashford, to co-operate in its creation and operation. The more radical suggestion, that a development corporation be set up, with direct powers from government to plan East Kent's economy before and after opening of the Channel Tunnel, was never seriously considered.

It is worth pausing to note that the gulf between British and French Channel Tunnel policy can be measured by this incident alone. The notion of the development corporation has been familiar in Britain since New Towns were first created in 1946, and, though modelled in this case on the Economic Development Committee of West Vlaanderen (centred on Bruges), the concept is equally consonant with French practice. It is, moreover, consistent with the thinking about local economic development of successive Thatcher governments. However, in the case of the Channel Tunnel, the political will to establish such a corporation in East Kent was wholly lacking, despite KCC support for the idea. Even the more limited proposal to institute a development agency came to nothing. As a result, policy in this important area has been without a consistent focus or direction, and development opportunities have probably been lost to Nord-Pas de Calais (and elsewhere).

There is, then, no comparison between the speeds of British and French policy responses to the Channel Tunnel: the Plan Transmanche was signed on 14 March 1986; the *Kent Impact Study: Overall Assessment* was issued in December 1987. Furthermore, these are very different documents, reflecting both a different ordering of priorities in Britain and France, and a different location of permissible Tunnel costs, and desirable Tunnel benefits. These two important differences will be considered in Sections 6.2 and 6.3.

6.2 Policy priorities

In both Britain and France, policy-makers have been concerned to ensure both that Eurotunnel is subject to sufficient control by public authorities throughout the period of its fixed-link concession, and that the economic benefits of the Tunnel scheme outweigh its costs. However, the balance struck between these twin concerns has been markedly different in the two countries. In France, as is evident from the *Eléments* submitted to President Mitterrand, economic development has been the primary policy objective, to which almost all other considerations are subordinate. In Britain, by contrast, the details of the planning regime to which

Eurotunnel is subject have been considered at very great length, and strategic economic development has been taken much less seriously than in France.

The difference in planning regimes is demonstrated by mechanisms used in Britain and France to authorise the Channel Tunnel scheme: in Britain, a lengthy hybrid bill procedure, which was entirely conventional, but nevertheless controversial; in France, a standard public inquiry, which was rapidly concluded, to the dissatisfaction of almost no one. Of interest here is not the degree of democracy embodied in these two procedures (which was discussed in Chapter 2) but the policy networks which lay behind and sustained them.

Elite negotiation of economic benefit in France

Again, it is the developed nature of French fixed-link policy networks which accounts for a large part of the observed difference in authorisation procedures in Britain and France. Just as the Nord-Pas de Calais Regional Council had its 'price' (PGC, strategic state investment in regional transport, tourism, and so on) for acquiescing in the French state's agreement with Britain to permit construction of a Channel fixed link (even though such a project was manifestly in the interests of the region), so, too, did implicated parts of Nord-Pas de Calais have a price which they extracted from the state and the Regional Council. The result is that even though some parts of the region — notably Calais, whose Chamber of Commerce (which runs the port) was a member of Flexilink — were as opposed to a fixed link as any of their British counterparts, this opposition was made known not through public but through private channels.

Details of different prices are readily found in the Plan Transmanche. Major opposition to a fixed link from the Calais Chamber of Commerce was bought off by the twin upgrading of the port itself, and of links to it (A26, improved access to the *rocade littorale*). Similarly, Boulogne was offered revitalisation of its substantial fishing industry in compensation for the decline its ferry industry would inevitably suffer following Channel Tunnel opening, and excellent road links to the north, south and east (*rocade littorale*, upgrading of the RN42 which links with the A26, A16). Furthermore, this pattern was replicated at the level of even the smallest communes affected by the decision to build, which in any case were set to profit from the massive increase in *taxe professionnelle* which Channel Tunnel construction would generate. Importantly, this increase, as the product of a construction project, would not be subject to *écrêtement* (the system whereby *taxe professionnelle* over a certain level is siphoned off for redistribution among neighbouring communes, which may not have secured the increase in local employment, but which nevertheless have to deal with some of its consequences). As a result, the communes around the Tunnel site will profit considerably from Eurotunnel, and are already devising substantial investment projects (such as a business centre at Fréthun and, more improbably, a *station balnéaire* (tourist resort) at Sangatte).

As was argued in Chapter 2, the public inquiry which came at the end of this negotiated process of mutual benefit was, therefore, a mere formality as far as local political elites were concerned. Furthermore, the programme of compulsory purchases which was triggered by promulgation (on 6 May 1987) of the Déclaration d'Utilité Publique (DUP), following successful conclusion of the French public inquiry, was also managed through elite accommodation, rather than through more diverse and public mechanisms. In this way, another class of potentially awkward local opposition — the farmers who would be expropriated to permit construction of the terminal site — was successfully managed. As the terminal site on the French side of the Tunnel is large — almost 900 ha if alterations to local roads and SNCF's land-take are included (compared with no more than 300 ha on the British side, split between three separate sites at Shakespeare Cliff (Dover), Cheriton and Ashford) — this was an important class of potential resistance. More than 80 farmers were involved, two of whom were projected to lose between 90 and 100 per cent of their land, three of whom were to lose between 75 and 89 per cent, four of whom were to lose between 50 and 74 per cent, and 14 of whom were to lose between 25 and 49 per cent.

Mechanisms to deal with this source of opposition date from before the abortive 1973 Channel Tunnel scheme. By declaring an area of possible development a *zone d'aménagement différé* (ZAD), public bodies are able to acquire the right to purchase saleable land at prices which do not reflect developer interest, but are instead pegged to prevailing prices in the land market.[4] This mechanism, which can be applied for a period of up to 14 years, was invoked at the start of the 1970s when the French state designated a ZAD of 1824 ha around the terminal site. On abandonment in January 1975, the state-owned share of the ZAD was some 150 ha. By keeping the ZAD in place, it was able to prevent alternative development in the area reserved for Channel Tunnel activities.

It was also able to minimise agricultural investment in the area, as land which is *zadé* is unlikely to be developed either by owner-occupiers (who in any case would find it difficult to secure bank loans for investment) or by absentee owners. Thus, when the fixed-link scheme was relaunched in the 1980s, the French government found itself faced with a farming community which was quite prepared to sell land, provided the terms were not disadvantageous. Represented in almost all cases by the Fédération Départementale des Syndicats d'Exploitants Agricoles (FDSEA), the local branch of the national farmers' union, land-owners and local farmers agreed to sanction elite negotiation of land sales.

It is worth noting that the body which took the initiative to represent farmers was a *syndicat*, and not a *chambre d'agriculture*, which was perhaps too closely managed by *notables* to secure local support. It is also not inconceivable that the prefect of the Pas de Calais prompted the FDSEA initiative: Chirac's Minister for Agriculture at the time, François

Guillaume, happened to be a former head of the national union, the FNSEA.

In these circumstances, *concertation* between buyers (the French state, Eurotunnel and SNCF) and sellers (represented by the FDSEA), was developed and successful. On the insistence of the FDSEA, a single land purchase office — a *bureau foncier unique* — was created by the three buyers, to allow negotiations to proceed on a one-to-one basis. As a result of these negotiations, a *protocole d'accord* was signed by the *bureau* and the FDSEA on 13 March 1987 to fix rules of land purchase. These rules were considered favourable to sellers, and allowed land acquisition to be concluded relatively rapidly. By April 1988, less than a year after promulgation of the DUP which triggers expropriation procedures, the full 890 ha of land had been acquired. In all, 400 separate properties had been bought, of which only 20 (5 per cent), accounting for less than 20 ha (2 per cent) of land, had been taken through judicial proceedings.

However, rapid land acquisition was not the only advantage of elite negotiation in France. A French law of 1962 requires that an expropriator who seeks to buy so much of a single holding as to endanger its viability must offer to buy the lot.[5] This requirement was enforced in 21 cases affected by *bureau* acquisitions. Of the 21 farmers, 16 subsequently left the land or the area, and five sought to continue farming locally. In these circumstances, the FDSEA was concerned to ensure that parcels of land acquired, but not needed for Tunnel installations, should be redistributed in the first instance between the five farmers who wished to continue to work in the area. This issue of *remembrement* was handled through *bureau*–FDSEA negotiations, along with the related problems of land drainage and access to local markets. To complement this extended programme of *concertation*, another *ad hoc* body, the Mission pour la Restructuration Foncière et Agricole du Calaisis (MIRFAC), was created in September 1988 by the regional and departmental councils, and the local *chambre d'agriculture*, to monitor resultant agricultural problems in the area around the terminal site.

In the context of the various deals that were struck in France, the issue of formal planning control was largely submerged. Measures such as the PGC were recognised as being of economic benefit to the coastal zone of Nord-Pas de Calais, centred on Calais, and although friction was caused by the entry into a crowded policy network of a powerful new actor — in the shape of the *coordonnateur grand chantier*, who drew his authority directly from the two prefects of the region — the principle of his delegated control of a number of important subsidiary aspects of the Channel Tunnel project was never seriously questioned. Similarly, the terminal site itself was declared a *zone d'aménagement concerté* (ZAC) — for the first time ever both by the state, rather than by a local authority, and to the benefit of a private (Eurotunnel) rather than a public corporation — but again the issue of control was considered less important than that of economic benefit. The very title, ZAC, suggests that development should be concerted, but it is clear that in dealing with a series of small communes, Eurotunnel will not be seriously constrained.

Indeed, in a number of areas, formal planning controls were virtually or actually ignored by the Tunnel project's developers — whether Eurotunnel or TML — with the overt or tacit connivance of the very authorities which were supposed to ensure their observance. Thus, despite having undertaken not to request the opening of gravel pits outside limits agreed with the French Ministry of the Environment, Eurotunnel soon sought to mine gravel at new pits in Sangatte and Wissant, and was granted authorisation for these works by the prefect of the Pas de Calais in an *arrêté* dated 9 March 1988. Criticised for this decision by a *député* at a round-table discussion 16 months later, the prefect, M. Dominé, explained that he had always tried to reconcile Eurotunnel's needs with those of the environment, and that, as he put it, he was driven by 'une obligation de résultat'. This, as the *Voix du Nord* noted at the time, was a surprising statement for a prefect to make, given that (in theory) his role is to uphold law, and not to engage in *ad hoc* acts of industrial promotion. However, as has been shown here, M. Dominé's statement is entirely consistent with the overall tenor of French fixed-link policy (*Voix du Nord*, 1989).[6]

Public debate of planning control in Britain

Formal planning control, of little concern to French policy-makers, was, by contrast, the central issue in British discussion of the Channel Tunnel project in the months following selection of CTG-FM in January 1986. Furthermore, it was an issue which was debated to a large extent in the public forum of Parliamentary Select Committee proceedings. This is not to say that there was no elite bargaining in Britain (nor that French concern with economic benefit had no salience in Britain). Indeed, behind-the-scenes discussions were an important part of the Select Committee process, and one of the main means by which authorisation of the Channel Tunnel project was secured on the British side.[7] However, the fact that these discussions took place in, or just outside, a public forum gave them a very different character to their French counterparts.

One of the reasons for this difference is that the extent of possible pay-offs for Kent was limited. Quite simply, Kent *qua* Kent had no need of a Channel Tunnel, did not really want one, and could certainly not be made to benefit from one to the extent that Nord-Pas de Calais could. In a French context, this would not have prevented it from seeking high levels of compensation. However, deeper institutional differences between the two political systems must also be cited to account for the full extent of policy divergence in this area. The comparative under-development of British political and administrative networking is a prime explanatory factor.

In conditions of institutional distance and separation, such as exist in Britain, the formal mechanisms of a planning system would seem to be accorded far more respect than they are in conditions of institutional interpenetration and complicity, such as exist in France. When all important decisions are taken by elite negotiation behind closed doors, correct procedure becomes a mere façade. When, however, decisions are made in accordance with known rules, respect for the rules becomes crucial. Certainly, such respect has been closely observed by all parties to the UK fixed-link debate.

Thus, perhaps the biggest issue to de debated during the Parliamentary passage of the Channel Tunnel Bill was the nature of the rules which would be embodied in its central planning regime. As was noted above, a merely pro forma clause was inserted in the bill, and all parties recognised that lengthy negotiation would be necessary before the relevant section could be inserted in a Channel Tunnel Act. The principle at stake — which no doubt seems arcane to French observers — was the extent to which the full planning permission of the pro forma clause should be modified to allow a degree of project control by local planning authorities (LPAs).[8] Having learnt some lessons from the 1970s project — which was thought to permit a reasonable, but not outstanding, degree of LPA control over construction works — and mindful of the minimal LPA control which is allowed by most private bills, planning officers at KCC and Shepway District Council in particular sought greater control over the 1980s project.

Specifically, they were concerned to secure control of Eurotunnel's construction activities (in particular, the movement of materials to and from the construction site), and of the design of its permanent developments (in particular, spoil disposal, landscaping of the construction site, and access to the Cheriton terminal). In each of these areas, LPAs managed to claw full planning permission back to equivalent outline planning permission. That is, Eurotunnel's consent was made equivalent to that granted under Part III of the Town and Country Planning Act 1971 (TCPA). Thereafter, the details of authorised works were to be submitted to the relevant LPA, which could only refuse, or impose conditions on, the granting of full planning permission on grounds specified in section 9, schedule 3 of the Channel Tunnel Act 1987.[9]

LPAs were reasonably happy with this regime, as it actually secured them gains which probably could not have been won under the TCPA. An example is Eurotunnel's undertaking to move as much bulk material by rail as possible, which could not be secured by means of a section 52 (of the TCPA) agreement, as section 52 usually applies only to the development of land, and does not extend to such matters as the transport of material to a construction site. It is important to note the dual dimension to this planning gain. Not only does the Channel Tunnel Act enable conditions (as understood in the TCPA) to be imposed on the developers of the Channel Tunnel project, but also it allows the imposition of undertakings which are not even recognisable as conditions by the TCPA.

This, then, is the kind of issue which preoccupied British fixed-link policy-makers in the months following selection of CTG-FM. Though debated by elites, it was a very different issue to that which was salient in Nord-Pas de Calais and, in addition, was handled in a very different manner to that which was adopted in Nord-Pas de Calais. Often, when British policy-makers sat down to negotiate, it was at the behest of a Parliamentary Select Committee, to which both sides had made representations, and by which both sides were now required to devise a mutually acceptable solution. In France, policy-makers never waited for licence to enter negotiation of contentious matters.

It would be wrong to state that British Channel Tunnel policy thus never looked beyond the detail of planning regimes in the months immediately after January 1986, even though such detail was a major concern. As has already been noted, the JCC commissioned the *Kent Impact Study* at this time. However, even this activity has to be set in context. The JCC established a number of study subgroups in 1986, of which the main ones were KIST, the Project Working Party (formerly Surface Works Working Party), the Planning Clauses subgroup, the Surplus Tunnel Spoil subgroup, the Traffic and Transportation subgroup, the Footpaths subgroup, and the Dollands Moor subgroup. As this list of titles indicates, economic development was far from being the main concern of the JCC in 1986, however highly the work of KIST was ranked.

It is generally agreed by all participants that the JCC was a valuable initiative, particularly in the period up to treaty ratification in July 1987. Indeed, the value of the JCC, and the importance of access to the network which its creation formalised, is corroborated by the strenuous efforts to which the Kent Association of Parish Councils (KAPC) went to secure representation on it.[10] Against any possible catalogue of achievements need to be set some (small) qualifications: that the JCC had no executive power; and that (therefore) it could only operate in conditions of co-operation and mutual respect, as is revealed by the experience of the short-lived London version of the JCC. This was established during the passage of the Channel Tunnel Bill in the second half of 1986, to discuss a proposed London rail terminal at Waterloo, and a possible west London railway line. It quickly disintegrated, however, when a number of Labour-controlled London boroughs (in particular, Southwark and Lambeth) demanded that meetings be held in public. If acceded to, this demand would have turned a species of rational administration into a platform for public posturing. It was therefore rejected by the DTp, the dissident boroughs walked out of the meeting in protest, and the Waterloo JCC was wound up.

However, the more important point that needs to be made about the JCC here is that it was so strongly concerned with planning controls, negotiated through the Select Committee hearings of the hybrid bill procedure. Moreover, this was not just an elite concern. Indeed, so strong was public disquiet about the Eurotunnel scheme, that a Channel Tunnel Complaints Commissioner, former diplomat Sir Donald Murray, was appointed by the JCC and installed in an office in Folkestone to receive complaints from the public about construction work, and to hear of failures to respect the provisions of the many planning regulations with which the Eurotunnel project had been surrounded in the UK.

This exaggerated British concern with full observance of the letter of the law may, as was noted above, be ascribed in part to institutional distance and separation, which breeds mutual suspicion, if not hostility, between levels of government. The distance and division which are characteristic of the British system have both positive and negative aspects, though, in the case of the Channel Tunnel at least, many people would agree that the negative outweigh the positive. In Chapter 5, the argument was made that the positive side is a high degree of co-ordination at the centre of government; and that the negative side is a low degree of co-ordination between central government and the rest of the governmental system, with which it is often required to collaborate in policy-making and implementation.

However, to this single positive aspect may be added another, which is that British central government, by being free of the ties which bind the centre to the periphery in France, is able to engage in policy initiatives which would probably be beyond the (institutionally constrained) capacity of the French state. An example in the case of the Channel Tunnel is land acquisition, which, largely because the amount of land involved was so much greater, was a more difficult problem in France, but which is nevertheless revealing of the comparative freedom of British central government to engage in policy initiatives.

It has been argued that the process of land acquisition was successfully managed in France, to the satisfaction of almost all parties. It was nevertheless an institutionally laden process which, though appropriate to the circumstances of the Channel Tunnel, could probably not have been avoided even if it had been inappropriate. It is simply difficult to believe that the project developers could have avoided *concertation*. In Britain, because central government is not structurally constrained, it has the freedom to engage in *ad hoc* initiatives which bypass lower levels of government. This happened in the case of land acquisition on the UK side, and undoubtedly resulted in better policy.

To say this is to pass (negative) judgement on the rules which regulate compulsory purchase in the UK, for in the case of the Channel Tunnel these were used only as a last resort, and better policy resulted from their virtual abandonment. The decision government took was to allow Eurotunnel to buy land — which was covered by residual compulsory purchase powers (vested not in Eurotunnel, but in government itself) — as and how it pleased. At no stage did acquired land become the property of Eurotunnel — instead, all land is leased by government as part of the 55-year concession — but the concessionnaire was allowed freedom to negotiate its purchase. The important difference this made was to release the process of land acquisition from some of the provisions of the Land Compensation Act 1973 (LCA), which regulates both the upper and the lower limit of compensation that must be paid for land by public bodies, but which regulates only the lower limit of compensation that must be offered by private developers. The result was that instead of being rigidly tied to the British doctrine that those with a just claim should neither lose *nor gain* from disruption, and in consequence being constrained to compensate up to, but not above, 100 per cent of market value, Eurotunnel was able to offer prices in excess of this limit, and thereby to smooth the process of land acquisition.

In the immediately affected villages of Newington, Peene and Frogholt (partly on the initiative of Folkestone MP and government minister, Michael Howard), it established the Three Villages Scheme, by which all villagers are offered the right to sell their houses to Eurotunnel at an agreed (and indexed) price at any time during the ten years following the concessionnaire's selection on 20 January 1986.[11] Beyond this, it engaged in some discussion with representative bodies, such as in particular the National Farmers' Union (NFU). However, the main concern of the NFU was not to challenge the arrangements by which Eurotunnel proposed to acquire land, but to question the statutory procedures embodied in the LCA. Thus, through *ad hoc* discussions with villagers and with landowners whose land was to be acquired, but without resort to the sort of corporate framework adopted in France, Eurotunnel purchased most of the land needed for construction to proceed. One or two landowners (not householders) preferred to deal with the Secretary of State rather than with the concessionnaire, and in these cases compulsory powers were used, but the vast majority of sellers dealt directly with Eurotunnel.

Thus, institutional distance allowed the British government more freedom in formulating fixed-link policy than was available to its French counterpart. On occasion this freedom was valuable, as is shown by Eurotunnel's process of land acquisition. Indeed, the point is emphasised by the difficulties which BR, a public company which cannot be released from the provisions of the LCA, is likely to face if and when it begins acquiring land for a Channel Tunnel rail link. In the main, however, institutional distance reinforced a British propensity to engage in discussion of rules, and to focus on correct procedure, to the detriment of consideration of the broader policy implications of a Channel Tunnel. In France, by contrast, broader policy implications were from the start the first concern of policy-makers at all levels of government.

6.3 Policy coverage

The final difference between British and French fixed-link policy in the construction phase of the present project is the locus of policy coverage in the two countries. In France, policy has sought chiefly to concentrate construction benefits in Nord-Pas de Calais. In the UK, it was constrained to disperse them throughout the national economy. This difference is evidently driven chiefly by economics, and by the comparative economic situations of the immediate hinterlands of the Channel Tunnel in Kent and Nord-Pas de Calais. However, what is interesting is the extent to which the resultant strategies adopted by policy-makers have reflected the preferred operations of their respective systems of government.

Containment strategies in France

One of the main reasons why the French government was interested in building a Channel fixed link in the 1980s (and in earlier years) was that it would aid regeneration of the Nord-Pas de Calais economy. Similarly, the main explanation for regional enthusiasm for the link (at a time — between 1981 and 1985 — when almost no French developer was committed to the project) is economic gain. Thus, containment strategies in France were one of the premises on which the project was relaunched in the 1980s, and were soon agreed by policy-makers at all levels of government. Significant resistance came only from those parts of the region — in particular, the mining basin — which felt that 'containment' might be read too narrowly, and thus fail to benefit any part of the region south-east of Lille (or even of St Omer).[12]

These concerns were perhaps unfounded. The main institutional device established to direct economic benefit in the construction phase in France, the PGC, is indeed focused chiefly on the coastal part of Nord-Pas de Calais, and particularly on the Calais area. It is thereby consistent with the Regional Council's fixed-link policy of redeveloping the coast by means of Channel Tunnel construction and operation. However, this focus is in many ways unavoidable. The direct benefits of Channel Tunnel employment, for example, cannot readily be spread far from the construction site, and it has turned out that some 50 per cent of the TML workforce in France has been recruited from the coastal towns of Dunkerque, Calais, Boulogne and St Omer. Similarly, many of the local difficulties which the PGC is designed to handle — in the realms of housing, transport, education and health — have a very local salience. In areas which can have a more distant regional benefit, however, the *coordonnateur* has sought to attend to the needs of the entire Nord-Pas de Calais region. This is particularly true of his attempted management of TML subcontracting. Thus, both local and regional benefit is being sought in Nord-Pas de Calais. This division can be used to structure analysis of containment strategies in France.

At the local level — the Tunnel construction site and its immediate environs, including Calais — a strategy has been developed to ensure both that local disruption is minimised and that local economic benefit is maximised. Much of this work is undertaken through classic French *concertation* procedures, whereby the Tunnel site *coordonnateur* does not actually provide services himself, but instead acts as a co-ordinator at the appropriate level of existing services, whether state, regional, departmental or communal. This is an important function, because, although the role of prefectures and sub-prefectures is to provide regional and subregional co-ordination of state services, this activity is not necessarily undertaken in a way that is appropriate to the very special circumstances of a construction site the size of that at Sangatte, where over 5000 workers were employed at the height of tunnelling activity. On the one hand, co-ordination of non-state services may not be easily handled by prefectures, not because the French state is incapable of influencing parallel or competing institutions, but because it may not be used to dealing with the sorts of problems raised by the Channel Tunnel. On the other hand, prefectures, and even sub-prefectures, may find it difficult to intervene at the appropriate level.

Thus, in many ways, the function of the Tunnel site *coordonnateur* at the local level is to establish a project base for the Eurotunnel scheme within the framework of existing institutions. Drawing his authority directly from the two regional prefects, the *coordonnateur* is in a fairly powerful position to undertake this activity, which is exemplified in the standard social realms of health, education and housing (large parts of which are departmental responsibilities since decentralisation). The other dimension to the *coordonnateur*'s activities is *concertation* with the private sector, particularly Eurotunnel and TML. Here, the *coordonnateur* cannot actually interfere with the commercial freedom of Eurotunnel and TML, but in France this is not necessarily an important constraint. Investigation of the linked policy areas of employment, education and training, each of which is a major priority for the region, illustrates some of the means by which fixed-link policy has been managed at the local level, whether through the offices of the *coordonnateur*, or through other mechanisms of *concertation*.

Employment is organised in an effective partnership between the Agence Nationale pour l'Emploi (ANPE), the national employment agency, and TMC (the French arm of TML). The role of ANPE is to seek recruits through a special office established at Sangatte and connected to the national ANPE network through the Minitel computer system. With TMC, it has negotiated joint preselection of candidates, though final selection remains, of course, the responsibility of TMC. Nevertheless, effective local co-ordination of TMC's employment needs is provided through the Sangatte office of ANPE.

Linked to the local office of ANPE is a Comité de Pilotage Educatif: the two organisations actually occupy neighbouring offices on route de Coquelles, Sangatte (though they have independent status). This is a direct initiative of the Tunnel *coordonnateur*'s 'Tunnel, Accueil, Développement' programme, which seeks to bring together various organisations in the Calais area. It is not, however, the only local body with an interest in education and training. In October 1987, the Regional Council decided to establish a training centre, the Centre Permanent de Formation, at Calais, at a cost of 15 million francs, to be financed jointly by the state, Regional Council, TMC and local chambers of commerce. Furthermore, a state–region protocol, Formation Transmanche, was signed on 16 November 1987, in which the state agreed to commit funds to regional training from the Fonds National à l'Emploi and the Fonds pour la Formation Professionnelle, which would be matched by the region and enable an application to the European Social

Fund to be made. At a meeting on 29 January 1988, the Regional Council subsequently adopted a Schéma Régional des Formations, which was the product of extremely wide discussions, and which represented a comprehensive training strategy for Nord-Pas de Calais.

Thus, in the realm of training, the *coordonnateur* is simply a further addition to already extensive policy networks. However, the project focus, mentioned above, that he is able to bring to the Tunnel scheme is a valuable addition to, and reorientation of, those networks. Moreover, close *concertation* with TMC has secured the contractors' participation in a training programme which meets not only the needs of Tunnel construction, but also of the post-construction period, when local unemployment is likely to rise sharply and to remain a major regional preoccupation. Thus, the Nord-Pas de Calais training initiative is important, and could generate a lasting improvement in local skills. Certainly, it is believed to be a great deal more sophisticated than its Kent equivalent.

At a regional level, the *coordonnateur*'s main activity has been to attempt to ensure that as many TMC subcontracts as possible stay in the regional economy. Again, the caveat about TML's commercial freedom needs to be made, and again it would seem that this is not a decisive constraint on the *coordonnateur's* ambitions.

Subcontracting has an extremely complex structure, reflecting the importance attached to regional retention of TMC orders. Indeed, so great was this importance held to be, that on 18 January 1986 a protocol was signed by the Regional Council, the Comité Interprofessionel Social et Économique (CISE) and the four tendering consortia, in which each undertook to 'favoriser l'utilisation des capacités et du savoir faire des entreprises régionales pour la sous-traitance des travaux et la fourniture de matériaux' (Conseil Régional Nord-Pas de Calais, 1986b, p.18). Though in reality no more than a formal undertaking, this protocol is seen as a valuable declaration of intent on the part of the winning consortium.

At the centre of the subcontracting network stands a Bourse Contact-Entreprise (BCE), established on an initial annual budget of 800 000 F to process subcontracted Tunnel work. The BCE is funded jointly by the Regional Council, the two departmental councils, the regional CISE, France-Manche and TMC, and is located in the same Calais building as TMC, in rue Mollien. The strength of its links with TMC may be judged in large part from this fact. Directed by a *groupe de concertation* presided by Noël Josèphe, president of the Regional Council, the BCE is able to ensure that information channels in the regional industrial sector are properly irrigated. At the start of construction work, in conjunction with chambers of commerce and professional organisations, it compiled a register of regional industrial firms that might be able to tender for subcontracting work. In compiling this register, a key role was played by the Fédération Régionale des Travaux Publics du Nord-Pas de Calais. Again, this was in many ways a regional pre-qualification exercise, which in no sense binds TMC's subcontracting, but which is nevertheless of clear benefit to regional firms. Linked to the work of the BCE is that of the Regional Council's Mission Transmanche, which very effectively provides market information through a regular bulletin, *Le Tunnel*, which contains information about subcontracts, employment, construction progress, and so on. The result of these many activities is that a steadily increasing proportion of subcontracts has been kept in the regional economy: by 1990, the figure was around 50 per cent by value.

Policy networks in Nord-Pas de Calais are thus developed and extensive, and operate at a variety of levels, of which the most prominent are local and regional. Two points may be made about these networks. One is that they are, of course, not without internal problems and conflicts. Thus, manifest tensions (in an era of 'decentralisation') have emerged between the Regional Council and the state over control of many of the services co-ordinated through the mechanism of the PGC. Similarly, difficulties in the sensitive area of TML's commercial freedom have sometimes been encountered. However, renewed *concertation*, and the evident desire of all parties to maintain good relations locally and regionally, have prevented these difficulties from becoming pronounced.

A second point is that French policy networks seem to be regionally structured to quite a considerable degree, permitting easy development of subnetworks — in this case, locally focused on the construction site, where the Tunnel *coordonnateur* has established a clear project base — but perhaps making inter- and multi-regional links difficult to institute. This suggests that the acknowledged success of containment strategies in Nord-Pas de Calais could perhaps not be translated into successful dispersal strategies, which have been the policy priority in Britain.

Dispersal strategies in Britain

At the outset, then, it is important to note that the economics of the Channel Tunnel in Britain and France has set policy-makers very different challenges in the construction period. Before judgement is passed on the British dispersal strategy, therefore, it is worth stating that such a strategy may be inherently difficult to mount. Equally, it is important to state that the British system would also probably find a containment strategy an awkward prospect, not having the developed intergovernmental networks that almost naturally exist in France.

Not sharing the French reason for building a Channel fixed link of precisely targetted *aménagement du territoire*, policy-makers in the UK simply had no clear strategy for public management of the project in the construction period. It was abundantly clear that few votes were to be won by promises of jobs and subcontracts for Kent, but it was not at all clear how the more attractive policy of dispersed benefits for the entire nation — particularly its depressed parts to the north and west — was to be developed. Furthermore, there was the problem of demonstrating to Kent exactly how it would benefit from such a major (and largely unwelcome) construction project.

It would not be fair to say that the policy response was to do nothing, though, equally, such an assessment would not be wholly inaccurate. From the start, the DTp, lead department throughout development of the 1980s scheme, had tried to talk to local authorities on as few occasions as possible, and this policy was not altered once a project had been selected in January 1986. It is true that the Department joined and chaired the JCC — it even claimed (as did many others) to have initiated it — but its strategy remained in many senses one of minimum necessary involvement. As far as the British government was concerned, the essential point about the Channel Tunnel was that it was a private-sector scheme. Others — such as the French, and British local authorities — might insist that it was impossible to deny the public-sector involvement in such a massive project, but the British government's line remained firm.

The political — or ideological — element to this stance must be noted. The government line was driven from Downing Street, and civil servants would seem to have departed from it without direct prime ministerial sanction. The peculiarly ambiguous policy response of the British (evident at many stages of the present project) no doubt stems in part from this tension at the heart of government between the ideological beliefs of successive Conservative governments, and the more pragmatic activities of civil servants. This is not to say that the DTp was anxious to develop extensive contacts with local authorities — clearly, it was not — but simply to note that ideology was a further constraint on already distant intergovernmental relations in Britain.

In the strong ideological climate of the 1980s, an institution such as the JCC looked decidedly anachronistic, and its main research output, the *Kent Impact Study*, was very much tainted by 1960s thinking and practice. However, its tri- or multipartism was no more developed than the 'toothless tripartism' which Hayward (1974) has derided. As has been noted before, the amount of policy contained in the *Kent Impact Study* was minimal, and carried little weight with policy-makers in either central or local government.

Thus, in Britain, distribution of construction orders has been left largely to the commercial decisions of Eurotunnel and TML. The single important exception has been a number of initiatives launched by the Department of Trade and Industry (DTI) under Lord Young. These include monthly publication, since January 1987, of a *Channel Fixed Link Newsletter*, which contains news about the Tunnel and other major European construction projects (such as the Danish Great Belt link), provides information about procurement seminars and possible procurement needs, and reproduces tender notices for TML subcontracts appearing in the 'S' Series of the EC's *Official Journal*. By the beginning of 1990, this newsletter had a print-run of 5500, of which some 2500 copies were sent direct to companies, the rest being distributed through DTI regional offices. However, these figures possibly understate the newsletter's true coverage, as DTI regional offices and chambers of commerce often reproduce articles, or even whole issues.

To complement this activity, and to raise its profile, the DTI in 1987 launched a series of regional procurement seminars, organised in conjunction with Eurotunnel and TML, which were designed to heighten British industry's awareness of the Channel Tunnel project, and of the sorts of order which might flow from it. The series was in part a response to regional anxieties, fed through the DTI's regional offices, and was therefore targeted in particular at areas, such as the West Midlands, which were thought to be performing badly (and which contained firms which might reasonably be expected to meet TML's needs). This series of seminars — in contrast to the subsequent DTI 1992 initiative — had the substantial benefit of offering tangible rewards to companies. TML representatives were present, and real contracts would evidently soon be issued.

Furthermore, through the newsletter, the DTI's Channel Fixed Link Section also operates a 'marriage bureau' for firms seeking joint ventures with partners from other Community countries. (Quite independently of anything organised by DTI, the British Embassy in Paris also operates a well-established joint ventures scheme.) This is simply an information service, and does not involve the DTI in marriage brokering, counselling or any of the other activities which are regularly undertaken by public authorities in France. The success of the scheme is hard to judge, partly because firms do not always want to reveal what they are doing. By the beginning of 1990, some 100 British and some 50 foreign companies had registered with the DTI, though only two known joint ventures had been organised through the scheme. This modest success probably understates the importance of the initiative, which has generated great interest among British companies, and which may only have a decisive impact in the comparatively long term. It also testifies to the difficulty of launching a multinational scheme. It is hard, for example, for the DTI to find French partners for British firms seeking joint ventures, because the main liaison in France is the BCE, which has a very small coverage and operates in ways which are alien to British practice. Nevertheless, in the increasingly European market in which British firms now operate, the marriage bureau is potentially the most significant aspect of the DTI's fixed-link initiative, and could prove to be of lasting importance.

The overall success of the DTI's dispersal strategy is also hard to monitor. At the end of 1989, some 80 per cent of TML subcontracts by value had apparently been retained by British firms (down from 90 per cent nine months earlier). Within this global figure, which is similar to that recorded in France, regional distribution of orders was less favourable to depressed parts of the national economy than had been predicted and desired.

At a subnational level, KCC has launched a training initiative which is largely imitative of that developed by Nord-Pas de Calais Regional Council, though nothing like as extensive in either coverage or support. Beyond this initiative, such matters as the site of TML's construction workers' village — it has been placed at Farthingloe, behind Shakespeare Cliff — have been a matter for normal LPA control, and not for special procedures.

Thus, a dispersal strategy has accorded well with the preferred operations of British policy-makers, which are to operate at arm's length from the commercial decisions of companies such as Eurotunnel and TML whenever possible.

6.4 Conclusion

This chapter has shown that French policy-makers were quicker than their British counterparts to respond to the decision to build a Channel Tunnel; that their policy priority was economic benefit, not planning control; and that their policy coverage was concentrated, not dispersed. These observed differences may be explained in part by economic, and in part by institutional factors. The important institutional argument is that, once economic differences have been controlled for, fixed-link policy becomes almost a caricature of Ashford's (1982) British dogmatism and French pragmatism thesis, with British policy-makers trailing a long way behind French.

In a way that would not be possible in France, British fixed-link policy-makers were able to maintain a degree of autonomy which may have accorded well with their preferred operations, but which cannot be said to have contributed to rational and efficient policy-making. Too often, separation caused their attention to be caught by legalistic matters, and the broader policy implications of the decision to build a Channel Tunnel were allowed to go unheeded. Crucially, no institutional mechanisms, such as are very present in France, existed to correct this failing and force broader issues on to policy-makers' agenda.

This is not a blanket critique of British institutions (and a bland celebration of French). As has been noted in this chapter and in chapter 5, French policy-makers' freedom of manoeuvre can be severely restricted by the vast policy networks in which they are inevitably implicated and can result in immobilism. This problem does not confront British policy-makers as frequently, and this can be an advantage. However, the negative consequence of British policy-makers' comparative freedom of manoeuvre is their comparative freedom to ignore uncongenial policy areas. In the case of the Channel Tunnel, this freedom proved to be destructive of efficient policy-making.

Notes

1. The general objectives of the *procédure 'grands chantiers'* (the full title of the procedure is '*grand chantier d'aménagement du territoire*') are set out in three decisions issued by the Comité Interministériel d'Aménagement du Territoire (CIAT) in July 1975, May 1980 and May 1982. For a full account, see Guillot (1988).
2. Subsequently, the Chirac government in 1987 decided to concede construction of the A16 (Paris–Amiens–Boulogne) to SANEF; and the current *Plan*

autoroutier français contains a commitment *à dix ans* to construct a *route des estuaires* to extend the *rocade littorale* as far as Bayonne. This latter remains, however, a somewhat vague commitment.

3. Le Syncrotron, a high-technology project written into the first (1984–8) *contrat de plan* signed by the French state and the Alsace region, and due for construction in Strasbourg, was subsequently unilaterally transferred to Grenoble by the French government (partly on technical and partly on political grounds). When the decision was challenged in the courts, the *contrat de plan* was found to have no juridical basis.
4. It should be noted that only land which comes up for sale is covered by this procedure; it does not permit expropriation (and it does not oblige the public body to purchase land).
5. The decision as to when this point is reached is taken by a judge.
6. The round-table discussion was in itself the sort of forum that only the French could have organised, bringing together three Ministers — of Transport (Delebarre), the Sea (Mellick) and the Environment (Lalonde) — two *députés*, two prefects and a host of representatives from Eurotunnel and TML (including their respective French presidents, Bénard and Essig), to discuss in open session the extent of Channel Tunnel progress, and means by which *concertation* might be improved (*Voix du Nord*, 1989).
7. See p.41.
8. The LPA is usually the relevant district council, except, in this case, where the source of aggregates, and the transport of materials to the construction site, are involved, in which two cases the LPA is the county council.
9. The two main grounds are that design be modified 'to preserve the amenity of the neighbourhood'; and that 'the development ought and could reasonably be carried out elsewhere' on the site mapped out by the Act. Each is difficult to uphold.
10. KAPC's problem was that the JCC is a grouping of statutory local authorities, and that KAPC itself is non-statutory.
11. The scheme was subsequently extended by Eurotunnel to cover 14 houses around the inland clearance depot at Ashford. The terms of the scheme are set out in the Commons Select Committee hearings.
12. It should be noted that resistance to agreed policies from the neighbouring Picardie region focuses on the TGV Nord route (through Lille rather than through Amiens), and is concerned to secure benefits in the operational, rather than the construction, phase of the Channel Tunnel. See Chapter 7.

7 The politics of Channel Tunnel operation in Britain and France

In its operational phase, the Channel Tunnel casts off the mantle of the world's biggest single construction project, and assumes the new character of key link in the transport infrastructure of north-west Europe. Its impact and importance then spread far beyond the local economies of Kent and Nord-Pas de Calais, to impinge on a series of regional and national economies ranged in concentric circles around the Tunnel itself. Policy-makers' concerns are similarly broadened, as they seek not simply to ensure a particular distribution of construction orders but to reposition regional and national economies in the new Europe which will be created by the EC's 1992 programme and opening of the Channel Tunnel.

However, local and regional constraints on public policy remain pervasive in the Channel Tunnel's operational phase. Although policy-makers seek to operate on a national scale in order to secure strategic economic advantage from the new transport system of which the Channel Tunnel is a central part, they are not even partially released from the local economic constraints presented in Chapter 4. When the clear identity of institutional and political structures, discussed in Chapter 5, is added to these constraints, it is evident that analysis in this chapter will be continuous with that developed in Chapter 6. Indeed, as far as the French are concerned, short- and long-term fixed-link policy-making have largely been coterminous, and even in the UK there has been a large degree of overlap between the two.

In these circumstances, the differences between British and French policy-making identified in Chapter 6 form the basis for analysis here. In planning the operational phase, as in planning the construction phase of the present Channel Tunnel project, French politicians and administrators have been quicker, more economically orientated, and more concerned to derive clear local benefit from the scheme than have been their British counterparts. As a result, fixed-link policy has continued to caricature Ashford's (1982) British dogmatism and French pragmatism thesis, and British policy-makers have continued to lag a long way behind their French counterparts. In analysing fixed-link policy in the operational phase of the present Channel Tunnel project, this chapter will first consider transport policy in Britain and France, before moving on to investigate regional policy responses, which aim to draw economic benefit from the new transport system which will be created by the Channel Tunnel.

7.1 Transport policy

The difference in British and French transport policy with respect to the Channel Tunnel is marked. In France, strategic planning was developed at an early stage in the present project. In Britain, market solutions were often favoured (or were used as an excuse for government inaction).

Strategic planning of transport infrastructure in France

The central instrument used to plan the transport infrastructure which would accompany Channel Tunnel operation in France was the Plan Transmanche, signed by the regional prefect and the Nord-Pas de Calais Regional Council on 14 March 1986, less than two months after selection of CTG-FM as fixed-link concessionaires. Its central commitments had been devised by the series of consultations which fed into the Regional Council's presentation to President Mitterrand at Lille town hall, which formed the basis for the Council of Ministers' rapid subsequent policy decisions.[1]

Into this plan was written a series of measures designed to ensure maximum regional and national benefit from the Tunnel scheme. As has already been noted, parts of the plan were chiefly intended to buy off opposition to a fixed link in port towns such as Boulogne and Calais, and to ensure that to some extent benefits from the Tunnel would be spread around the Nord-Pas de Calais region (and particularly its coast). However, there was also a clear strategic dimension to the plan, which is not always easy to separate from local compensatory measures, but which is nevertheless central to it.

Negotiation of this strategic dimension was conducted through the *contrat de plan* system, enshrined in the law of 29 July 1982 which reformed the French planning system by giving it a regional dimension. Under the new mechanism, the state agrees a regional plan with each of the 22 French regions, both as a means of attending to regional economic problems and as a means of developing a national plan. Exceptionally, this mechanism was activated to deal with the special case of the Channel Tunnel, and the full panoply of consultation was undertaken, feeding local views to the Regional Council through both political and socio-professional local networks, and feeding regional views on to the state through equivalent contacts at the regional level (prominent among which was the Regional Council itself). The whole was then brought together in a range of national plans, among which the *schémas directeurs* for roads and rail are particularly relevant to this study. Consistent with these *schémas directeurs* are the Plan routier Transmanche, and SNCF's plans for TGV Nord.

The rhetoric in which these plans are steeped needs to be read with great caution, if strategy is to be separated from compromise and accommodation to local interests. Thus the declaration that 'Le gouvernement français a décidé le lancement du plan routier Transmanche, destiné à désenclaver le littoral' (Direction Départementale de l'Équipement du Pas-de-Calais, 1988), tells only part of the story by which this particular plan was developed. Nevertheless, it is a reasonably accurate description of the likely result of the series of measures which it contains. Shown in Figure 7.1, these are:

(1) Completion of the north-east to south-west (non-toll) *rocade littorale*, stretching from the Belgian border to Boulogne-sur-Mer, which is designed to improve access to and from the entire coastal strip of north-west France, and which will link not only with the Belgian motorway system, but also with routes to the South-West of France, such as a *route des estuaires*, which bypass Paris.
(2) Completion of the north-west to south-east A26 toll road, designed to link Calais, Arras and Reims, and thereby to give access to south-east France, also bypassing Paris.
(3) Completion of the A16 toll road, designed to provide a direct route from Boulogne (via Amiens) to Paris, and an alternative route to Paris to the A26/A1 (via Arras) from Calais, and the A25/A1 (via Lille) from Dunkerque.

Plans for TGV Nord followed a slightly different timetable, because part of the network, known as PBKA (Paris–Brussels–Cologne–Amsterdam; K for Köln) was being planned independently at the beginning of the 1980s, and would (it is said) have been undertaken with or without a fixed link. In these circumstances, all that was needed was to agree a spur from Lille (on the route from Paris to Brussels) to the Channel Tunnel mouth at Fréthun to turn PBKA into PBKAL (L for London). This made the choice of a route due north of Paris, paralleling the A1, to Lille, and due west from Lille to Fréthun, relatively straightforward, and facilitated rejection of the competing claims of Amiens. These were apparently strong, as Amiens sits astride the route as the crow flies from Paris to Fréthun, and would offer a saving of distance and time over the alternative route through Lille. However, on grounds of *aménagement du territoire* (and as a result of intensive Nord-Pas de Calais lobbying through its excellent links into the centre of French government), the Lille route was chosen (and had, indeed, been reserved in local plans since the failure of the 1970s Tunnel scheme).

Two separate study groups had been set up to consider these various proposals. One was an international group of railway and government officials from the four countries implicated in the PBKA scheme, which was concerned to consider the main outlines of a possible international high-speed rail network. (The British government and BR sent observers

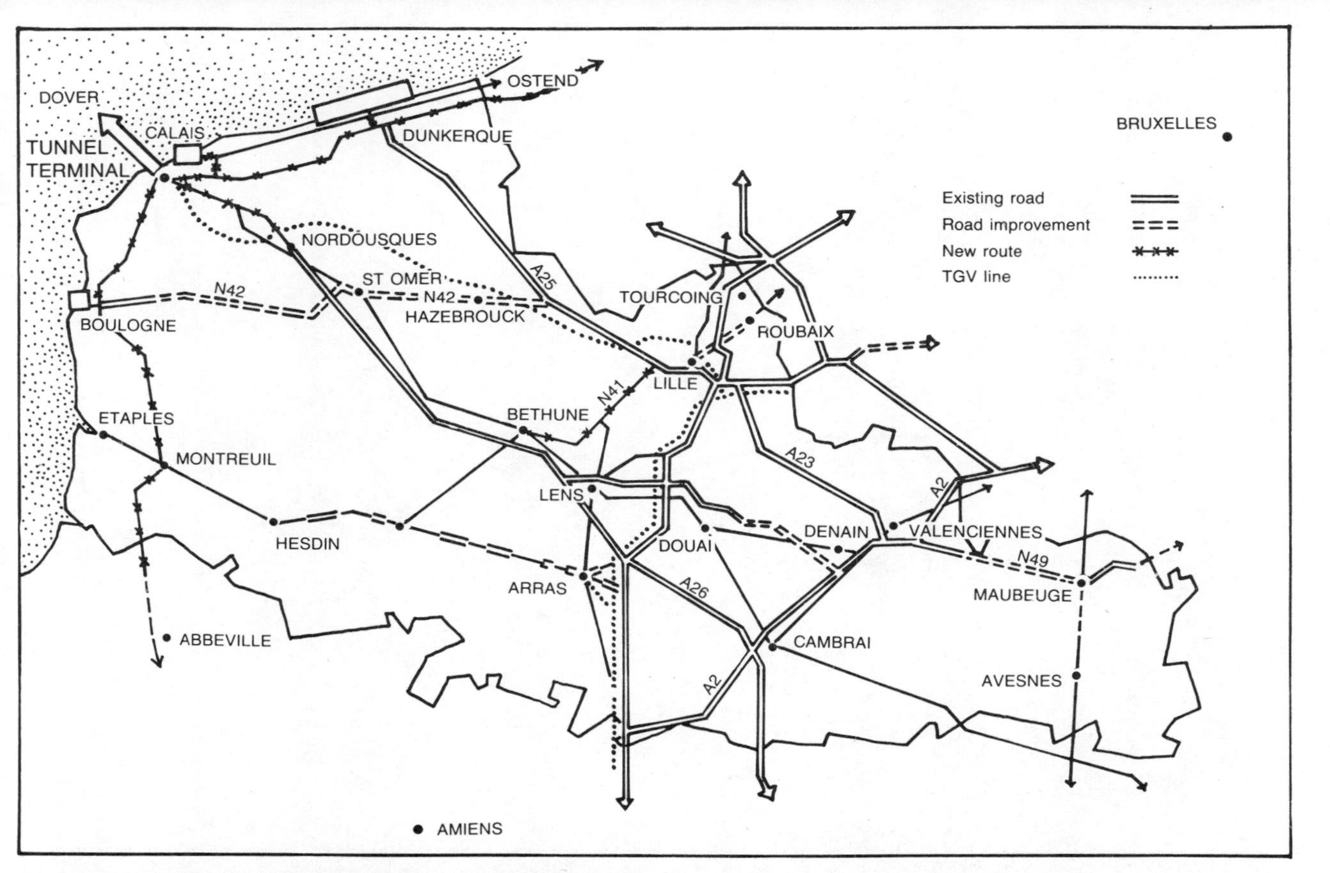

Figure 7.1 *Transport infrastructure in Nord-Pas de Calais*

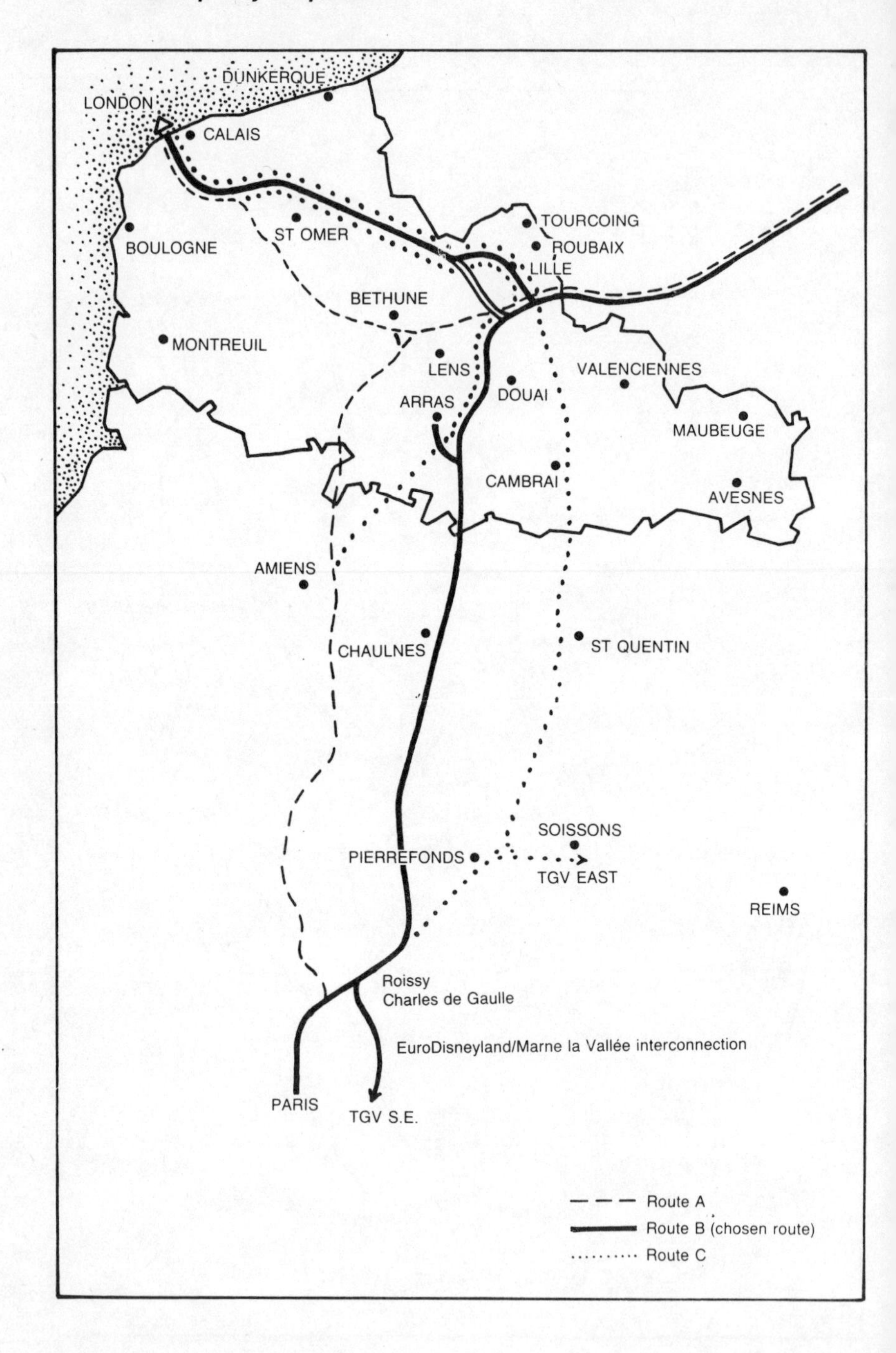

Figure 7.2 *TGV Nord alternative routes*

Table 7.1 *Cost of alternative routes for TGV Nord (in millions of francs at June 1985 prices)*

Route	
Plaine des Flandres route	10 805
South of Lille route	12 405
Extra cost of line through centre of Lille	920
Extra cost of Roissy link	860–990
Extra cost of Amiens alternative	290

Source: Conseil Régional Nord-Pas de Calais (1987)

Table 7.2 *Traffic forecasts for competing TGV Nord routes (in millions of passengers per year)*

Service	Plaine des Flandres	South of Lille	Lille centre
Paris–London	9.3	9.5	9.0
Paris–Brussels	5.7	5.7	5.7
London–Brussels	5.6	5.5	5.6
Paris–North France	8.8	8.8	8.8
London–North France	1.9	1.9	1.9

Source: Conseil Régional Nord-Pas de Calais (1987)

to this group once a fixed link had become a definite possibility in the mid-1980s, and became a full participant once CTG-FM had been selected.) The other was a domestic study group, directed by SNCF, which was concerned simply to investigate the technical and economic feasibility of TGV Nord and, if a positive conclusion was reached, to judge between competing routes.

Concluding that TGV Nord was likely to be both technically feasible and economically profitable, the SNCF team initially considered two possible routes. One was known as the 'Plaine des Flandres' route, the other known as the 'south of Lille' route. However, in response to specific requests (from local authorities in particular), it also subsequently investigated three further alternatives: one through the centre, rather than to the south, of Lille; one with a deviation just north of Paris to serve Roissy-Charles de Gaulle airport; one to take the main line north of Paris closer to the city of Amiens. Possible routes for TGV Nord are shown in Figure 7.2. At June 1985 prices, the various alternatives were priced as shown in Table 7.1, with corresponding traffic forecasts for the three Lille alternatives estimated as shown in Table 7.2.

The procedure chosen to decide between these competing routes on wider grounds of *aménagement du territoire* followed that adopted earlier in consideration of the route of the TGV Atlantique. A special commission, under the presidency of M. Rudeau, a senior civil servant, was formed to consult with politicians, socio-professional interests, and chambers of commerce and agriculture, before settling on a definitive route. Only when this had been done were the formal procedures of a public inquiry activated.

Of particular interest is the decision taken over the route through the centre of Lille. SNCF's proposal, to take the TGV round the southern outskirts of the city and build a station on a green-field site, was considered by the city fathers to be less favourable to economic development than a route through the middle of the city, stopping in the city centre itself. As SNCF was unwilling to pay the extra cost of this option (but was prepared to accept the lower volumes of Paris–London traffic projected to use it), specific funding for this part of TGV Nord was provided jointly by state (50 per cent) and regional/local (50 per cent combined) bodies, acting in concertation. On this basis, a range of development proposals was subsequently devised.[2]

Taking this set of measures as a whole, it is evident that French policy responses to the strategic challenges posed by the Channel Tunnel were rapid in the crucial area of related transport infrastructure investment. With this series of questions settled, it was possible for policy-makers to turn their attention to regional economic development issues. The key contrast with the situation in Britain is demonstrated most vividly at this point.

Market solutions in Britain

For two reasons, the notion of market solutions in Britain is inaccurate. First, a series of road improvements was written into the Channel Tunnel Act 1987, and will be financed by government in the usual manner. Second, responsibility for rail services to the Channel Tunnel, including those on the key route from Cheriton to central London, has been vested in British Rail, a state-owned company which is subsidised through general taxation. Nevertheless, the notion remains valid because, in approach and intent, the British government has sought wherever possible to adopt market solutions to the infrastructure needs of the Channel Tunnel.

Certainly, its approach has been less strategic than that of the French government. Where in France relatively ambitious plans have been developed not only to provide a choice of routes by road to and from the Nord-Pas de Calais coast, but also to link these into a strategic national plan for roads, in Britain all upgradings of the road network have been concentrated in Kent, and have been either necessary improvements to deal with projected Channel Tunnel traffic or compensatory measures to undermine dissent in port towns and other disadvantaged parts of the county. In the first category are completion of the M20 'missing link' between Maidstone and Ashford, upgrading of the A259 near Folkestone, of the A2070 to the south of Ashford, and of the A260 to the north of Cheriton, and a number of minor improvements to county roads near the Tunnel terminal. In the second are dualling of the A20 between Cheriton and Dover, and of the A299 (Thanet Way) from the end of the M2 to Ramsgate, and upgrading of a section of the A256 between Ramsgate and Dover. These routes are shown in Figure 7.3.

Strategic investment in, for example, a British *rocade littorale* stretching from Kent to Hampshire has been left at the study stage by the Department of Transport (1990), which takes the view that proven demand for such a route is not sufficient to justify the investment (and is not prepared to build the road as a stimulant to economic development). Some upgrading of existing routes (A27/A259) between Folkestone and Portsmouth is, however, included in current plans (ibid.).

The most striking and visible difference between British and French approaches to related transport infrastructure is, however, in rail investment. In contrast to the rapid French decision to take TGV Nord to the Tunnel mouth, British decision-makers continue to debate rail routes to the British end of the Tunnel, and are apparently no closer now to a decision than at any time since 1987. Part of the explanation for this difference is to be found in the economic constraints on policy-making presented in Chapter 4. At its most basic, the difference is that it is simply a great deal more difficult to thread new train paths through rural, suburban and populous Kent, than it is to build a new TGV line through Nord-Pas de Calais, which has a high population density (316 people per square kilometre) by comparison with the rest of France (101 people per square kilometre), but which remains significantly less densely populated than Kent (402 people per square kilometre). However, an explanation also has to be sought in the operations of British and French institutions.

It has already been noted that the rapid tripling of the cost of a British Channel Tunnel rail link effectively killed the 1970s Tunnel project. With this experience at the back of policy-makers' minds, it was perhaps not surprising during debate of the 1980s project to find both that British Rail was prepared to tell Parliament that it could cope with Channel Tunnel traffic on existing lines in the first years of operation, and to discover that no one sought clarification of the precise meaning of this phrase.[3] The issue was simply too contentious to be explored by fixed-link enthusiasts, and was inexplicably underexploited by fixed-link critics. What is surprising to French observers is the piecemeal fashion in which rail links to the Channel Tunnel were debated in Britain.

In France, a rail corridor between Lille and the Nord-Pas de Calais coast west of Calais had been reserved on local plans since abandonment of the 1970s project, and, as was noted above, was used in the 1980s as a central part of a classic strategy of *aménagement du territoire*. In the UK, no such vision was developed. Instead, Tunnel rail links were, and remain, locked into a debate about routes and funding which shows little prospect of resolution.

It is important to note that the key issue on the British side is not simply speed (of travel across Kent). Indeed, more critical is capacity on

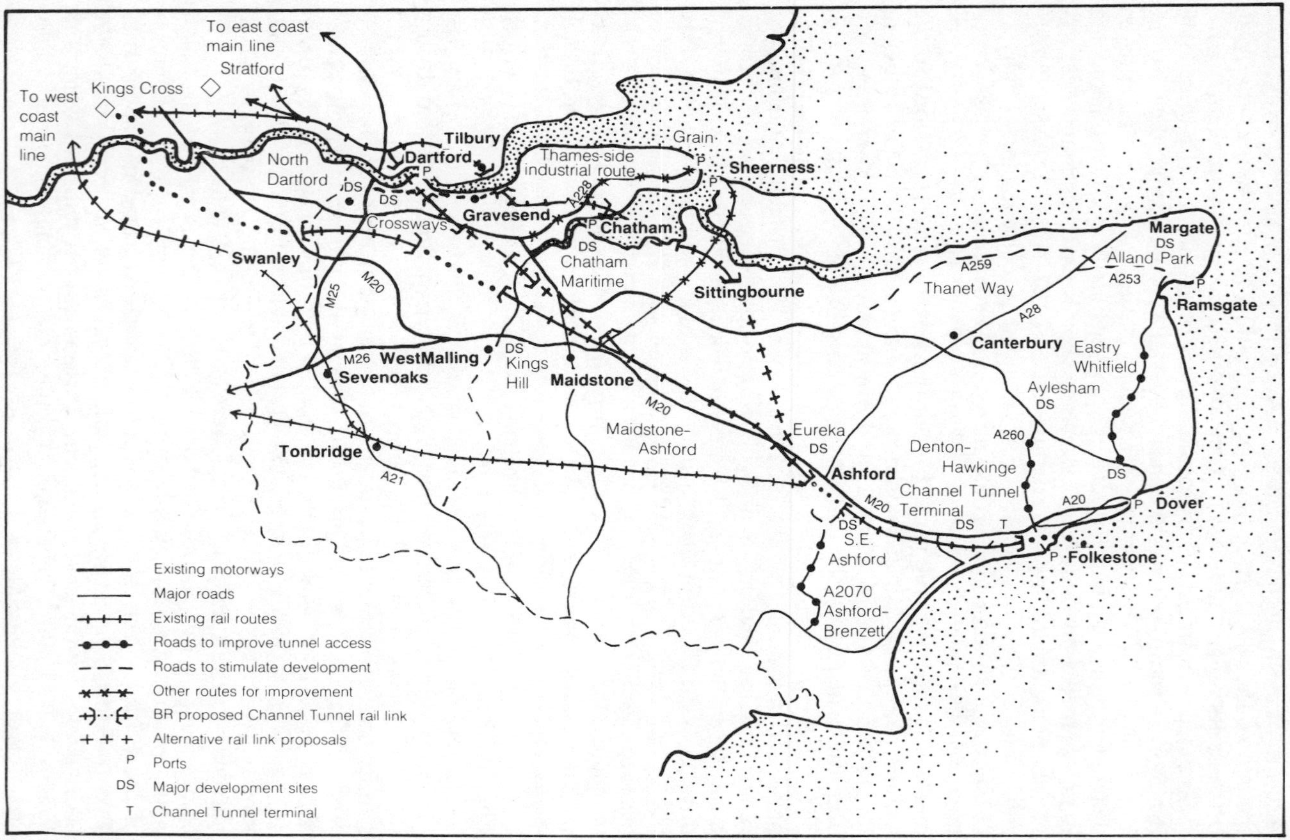

To east coast main line
Stratford
To west coast main line
Kings Cross
Tilbury
Dartford
North Dartford
DS
P
Thames-side industrial route
Grain
Sheerness
A228
Gravesend
Crossways
Chatham
DS
Chatham Maritime
Swanley
M25
M20
Sittingbourne
Margate
DS
Alland Park
A259
A253
Thanet Way
Ramsgate
A28
Canterbury
Eastry
Whitfield
Aylesham
DS
M26
WestMalling
DS
Kings Hill
Maidstone
Sevenoaks
M20
Maidstone-Ashford
Eureka
DS
A260
Denton-Hawkinge
Tonbridge
A21
Ashford
Channel Tunnel Terminal
DS
T
A20
Dover
Folkestone
S.E. Ashford
A2070 Ashford-Brenzett

Existing motorways
Major roads
Existing rail routes
Roads to improve tunnel access
Roads to stimulate development
Other routes for improvement
BR proposed Channel Tunnel rail link
Alternative rail link proposals
P Ports
DS Major development sites
T Channel Tunnel terminal

Figure 7.3 *Transport infrastructure in Kent*

Network South East, and consequent reliability of the service which British Rail is able to offer. It quickly becomes clear that this is actually the basic issue in discussion of British rail links to the Channel Tunnel, more fundamental than speed, which is no more than a refinement by comparison. A further issue (which is also more important than speed) is access to the European high-speed network, which is crucial not only for London and south-east England, but also for the whole of Britain. Thus, the three issues of capacity (hence reliability), access and speed (in that order) present themselves on the British side, and arise in discussion of both passenger and freight services through the Channel Tunnel.

The question of capacity, dependent as it is on traffic forecasts, is open to interpretation. On the one hand, it is necessary to determine the likely future demand for existing services on and through Network South East as a means of assessing residual Tunnel-related capacity. On the other, it is necessary to gauge the likely future demand for rail services through the Channel Tunnel. As regards the first issue, few would view British Rail's estimate of residual Channel Tunnel capacity on Network South East of some 15 million passenger trips per annum as an understatement of likely commuter demand. Similarly, British Rail's suggestion that 29–32 freight trains per day could be accommodated to give 'reliable capacity' across Kent is unlikely to be readily exceeded, particularly as other constraints — such as maximum throughput at Dollands Moor Inspection Facilities (which potentially all outbound Tunnel rail freight traffic must use) of 35 trains per day — come into play as soon as this limit is raised (Steer Davies & Gleave, 1989, p.16). The second issue is, however, more controversial.

The usual macro-economic uncertainties — concerning growth, inflation, and the relative prices of raw materials, for example — apply to both passenger and freight forecasts. In addition, each is subject to specific micro-economic unknowns. In the case of passenger forecasts, the key debate concerns traffic generation. Investigation of the experience of TGV Sud-Est, opened between Paris and Lyon in two stages in 1981 and 1983, has shown that this can be substantial. In global terms, the number of journeys on this route grew from 12.24 million in 1980, to 18.36 million in 1984, an increase of 50 per cent (compared with a predicted 2 per cent increase in absence of the TGV). Of the 6 million extra journeys, 2 million (33 per cent) are estimated to have transferred from Air Inter, the domestic airline, 1.1 million (18 per cent) to have transferred from roads, and 2.9 million (49 per cent) to be new business on the route (Laboratoire d'Economie des Transports/Interalp, 1986). Thus, in the second year of full operation, traffic generation of almost 25 per cent was witnessed on the new service, and total 'induced' traffic (defined as road plus trip generation) stood at some 33 per cent.

SNCF has used this experience as the basis for its estimate of Channel Tunnel passenger traffic, and consequently arrives at higher traffic forecasts than does British Rail, which assumes that British drivers are unlikely to switch to rail because with or without the Tunnel they will want to use their cars once they have crossed the Channel (Steer Davies

Table 7.3 *Summary of international rail traffic forecasts for the Channel Tunnel*

Year	Forecast	Passenger (million trips p.a.)	Freight (million tonnes p.a.)
1993	BR	13.4	6.1
	Eurotunnel	14.0	7.2
	SNCF	16.5	7.2
2003	BR	17.4	7.0
	Eurotunnel	24.7	12.2
	SNCF	21.4	10.6
2013	BR	21.2	7.7
	Eurotunnel	28.9	18.1
	SNCF	26.2	13.4
2023	BR	25.9	8.5
	SNCF	31.9	16.4

Note: In all cases figures do not allow for the likely generative effect of a high-speed link in Britain, but do include TGV operation in France, Belgium and beyond.
Sources: Eurotunnel (1990a); Steer Davies & Gleave (1989)

Table 7.4 *Eurotunnel's traffic forecasts (millions of passengers/millions of tonnes per annum)*

	1993*				2003				2013			
	1990	1989	1988	1987	1990	1989	1988	1987	1990	1989	1988	1987
Passenger												
Shuttle	14.6	15.8	15.3	13.2	19.9	22.9	21.5	18.1	25.0	29.0	27.4	20.5
Rail	14.0	13.6	15.4	16.5	24.7	21.0	19.8	21.4	28.9	25.0	22.4	26.1
Freight												
Shuttle	9.0	9.0	8.1	7.5	14.6	14.2	12.2	10.5	19.9	19.2	16.2	13.2
Rail	7.2	6.4	7.4	7.3	12.2	10.6	11.4	10.6	18.1	15.6	16.4	14.6

* 1993 figures are for a full year of operation
Source: Eurotunnel (1990a)

& Gleave, 1989, p.11): see Table 7.3. As can be seen from Table 7.4, Eurotunnel's forecasts have undergone substantial modification (as cost estimates have increased).

As regards freight, a crucial factor is the extent of diversion to rail that can be expected to follow a marked increase in the length of possible rail journeys to and from UK destinations. Working on the global assumption that road is more competitive than rail for freight journeys of less than 300 km, and that above this figure the reverse is true, it is possible to argue that substantial increases in rail's share of the freight market are likely to be the consequence of Channel Tunnel operation.

Certainly, BR can be expected to increase its current freight market share. On international routes, this is no more than about 1 per cent, which compares very unfavourably with the 16–20 per cent enjoyed by Continental railway companies. Domestically, it is around 8 per cent, but still below that of continental railways, which is typically higher than the 16–20 per cent domestic share (Vickerman and Flowerdew, 1990).

It will be seen from these sets of figures that maximum capacity on Network South East is reached either on, or very soon after, Channel Tunnel opening in June 1993. In the passenger market, SNCF estimates that 15 million passenger trips per annum will be exceeded in 1993; Eurotunnel and BR that it will be exceeded very soon thereafter. In the freight market, even BR's conservative 1993 forecast of 6.1 million tonnes per annum translates into 28 trains per day (Steer Davies & Gleave, 1989, p.15), and thereby places the network at just below total operating capacity on Tunnel opening. Eurotunnel and SNCF estimates would use all freight capacity from the start of Channel Tunnel operation. In both sectors, therefore, securing sufficient capacity on the rail system between central London and the Channel Tunnel is likely to be a major problem in the mid-1990s. The problem cannot not now be resolved before about the year 2000, and could set rail links to the Tunnel on a vicious spiral whereby unreliable services cause passengers and transporters to abandon rail, and government consequently to refuse to sanction investment in an underperforming service.

The second issue of access to the rail network which will pass through the Channel Tunnel is one which is clearly dependent to some extent on capacity on Network South East, for access throughout Britain will be diminished by constraints here. However, it also stretches well beyond this to address the adequacy of Britain's national rail network in an era of substantial rail investment, and projected increased rail use, throughout Europe. Two main problems confront policy-makers. First, the British loading gauge is different to the UIC or Berne gauge used throughout most of continental Europe, thus special trains will be needed to provide both passenger and freight services. Initially a core passenger service will only be provided between London and Paris or Brussels, with very limited through services from Manchester, Birmingham, Edinburgh and Leeds; the freight service depends on the development of regional depots and on new wagon technology. Whole sections of the country will thus be effectively denied direct access to the European rail network (Figure 7.4). Second, even if these changes are made British railways do not permit very high-speed operation (as is attained by the French TGV on dedicated track), such that Britain is left with what is increasingly being seen as a second-rate rail system. Each problem is now the focus of regional discontent, led by such bodies as the Royal Town Planning Institute (1990), and the North of England Regional Consortium (see, for example, Gunnell, 1988).

The necessary condition for high- and very high-speed operation on the British rail network is, first, that lines be electrified, and, second, that dedicated lines be built on the model of new TGV links in France (and,

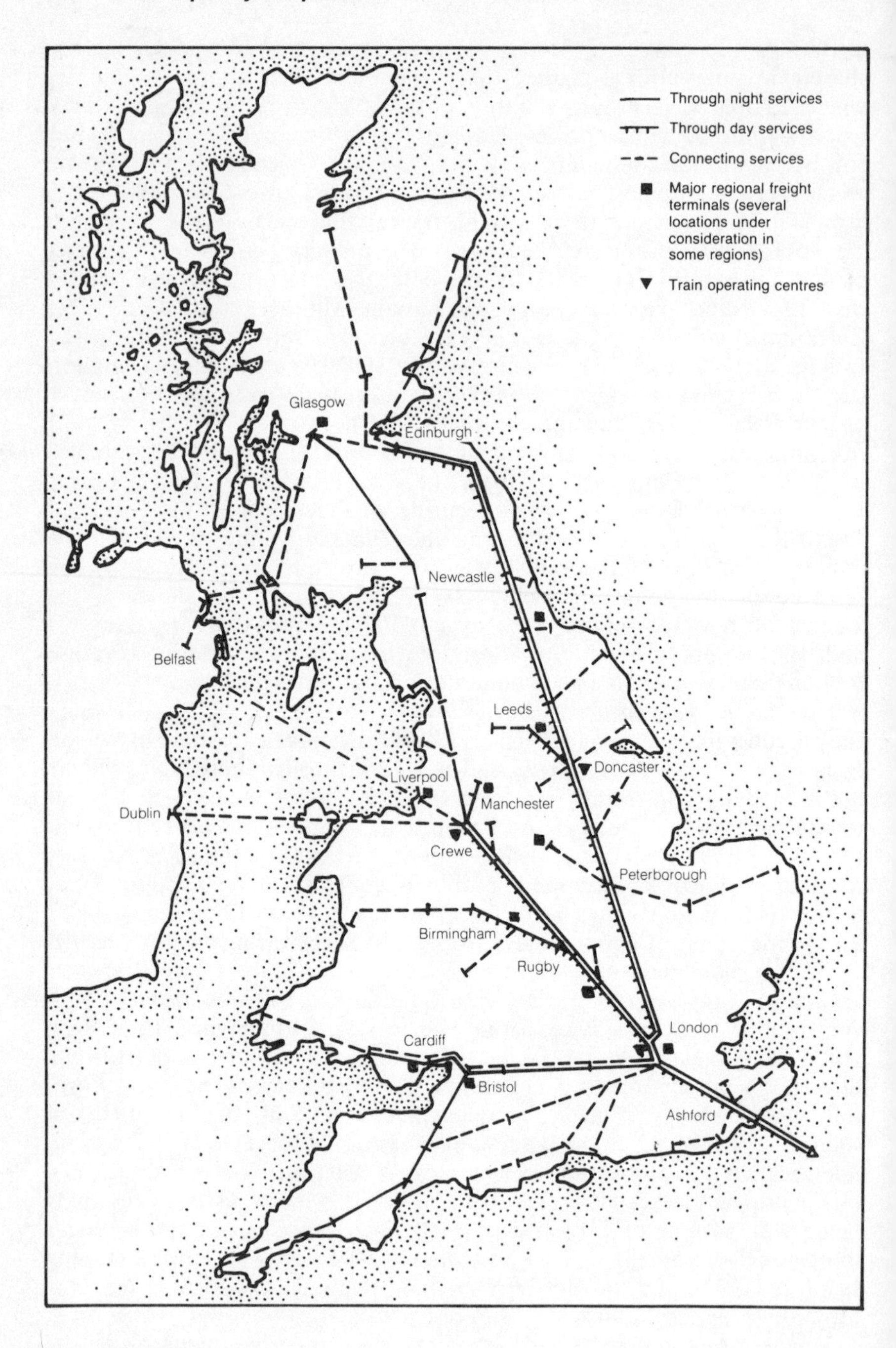

Figure 7.4 *Proposed international rail services*

increasingly, elsewhere in north-west Europe). To understand why the second of these conditions is unlikely to be fulfilled, a number of factors need to be understood. First, distances between major population centres in the UK are typically smaller than those found in many other European countries. Second, this concentration of the rail network is reinforced by the absence of long-distance international rail services because of the historic separation of the UK network from that of continental Europe. Third, despite the relatively densely developed network, the topography of lowland Britain is such that even early railways could be laid out, without major engineering problems, to relatively easy gradients and large radius curves. Fourth, the recent history of railway development in Britain has been dominated by the rather different attitude to railway investment and subsidy taken by successive British governments (of both major political persuasions) from that taken by their Continental counterparts.

7.2 Regional policy responses

What is clear from this review of British and French transport policy with regard to the Channel Tunnel is that French policy has been developed and agreed a great deal more quickly than British. The many qualifications that have been made to a simple — even simplistic — critique of British policy based on a straightforward comparison with French may now be set aside in order to investigate the implications of this basic difference for development of regional policy responses.

Aménagement du territoire in France

A phrase which has appeared at many stages in this account may now be set at the centre of analysis. *Aménagement du territoire* is a concept which was developed in France after the Second World War, and which, though not directly translatable, may be rendered in English as 'regional economic development planning'. It has been a major feature of French policy responses to the Channel Tunnel.

As has been noted before, strategy in France needs to be carefully separated from the compromise and accommodation which are standard features of the interpenetrating French state. In this section, only genuine measures of *aménagement du territoire* will be discussed. Thus, road links to the port of Calais will not be included, but the decision to route TGV Nord through the centre of Lille will be.

Indeed, this policy is the central concern of this section. On the basis of it, local and national policy-makers have sought to redevelop the Lille metropolis, and thereby to reinvigorate the entire Nord-Pas de Calais region. Two initiatives in particular require mention: the first, Euralille Métropole, being focused on the city of Lille itself; and the second, Projénor, being centred on the Nord-Pas de Calais region, but stretching well beyond its borders. Each is a *société d'économie mixte* (SEM), created in 1988.

Euralille Métropole is the joint initiative of the Mayor of Lille, Pierre Mauroy, and of a number of public-sector bodies, among which the role of the Caisse des Dépôts et Consignations (CDC) was particularly important. Its shareholders are major financial institutions — CDC, Crédit Lyonnais, Banque Indosuez, Banque Scalbert-Dupont, and Banque Populaire du Nord — and its associated members are the Chambre de Commerce et d'Industrie de Lille, Roubaix et Tourcoing, and SNCF. Its president is M. Jean Deflassieux, former president of Crédit Lyonnais. The role of Euralille Métropole is to plan and design an international business centre around the TGV Nord station, due to open in June 1993, as a means of attracting private investment for its development (Ville de Lille, 1989). Thus, this is essentially a pump-priming exercise, aimed at offering a fully serviced trade centre to potential private-sector investors.

Projénor is similarly a public–private initiative, and brings together a number of the institutions involved in Euralille Métropole. Again, the role of the CDC was crucial in getting the project off the ground, though interestingly (to students of decentralisation and *déconcentration*), it was the regional director, not a member of the national agency, who was the prime mover in project development. Projénor's shareholders are the CDC (which holds 50 per cent of the stock), Banque Nationale de Paris, and Crédit Lyonnais. Associate members are Crédit Agricole, Midland Bank, Crédit Communal de Belgique, Société Régionale des Caisses d'Epargne (SOREFI), Société de Développement Régional (SDR), the two regional reconversion agencies, SODIKERQUE and SODINOR, the regional chamber of commerce, Eurotunnel, SNCF, its subsidiary, SCETA, and Société des Autoroutes du Nord et de l'Est de la France (SANEF), the regional motorway concessionaire. The president of Projénor is M. Jacques Sallois, previously an employee of DATAR, and director of the CDC's external relations.

The main purpose of Projénor is to pump-prime on a regional (and pan-regional) scale by developing projects both in Nord-Pas de Calais and in its neighbouring French, British and Belgian regions. By conducting feasibility studies, and by writing technical and financial reports, it aims to complement the activity of Euralille Métropole, and to attract capital to a part of north-west Europe which, historically, has been underdeveloped, but which as a result of its close association with the Channel Tunnel may be able to reshape its fortunes. The very small size of its capital base — 10 million francs — is a warning against reading too much into the existence of Projénor, but it does bring together an interesting mix of public and private institutions, and could fulfil an important role in the *valorisation* of the Channel Tunnel in northern France at least (though perhaps not elsewhere: almost all its present projects are in Nord-Pas de Calais).

Along the length of the Nord-Pas de Calais coast, a tourism initiative, the Mission Côte d'Opale, has been launched in an attempt to profit from the heightened visibility of the region which results from the Tunnel

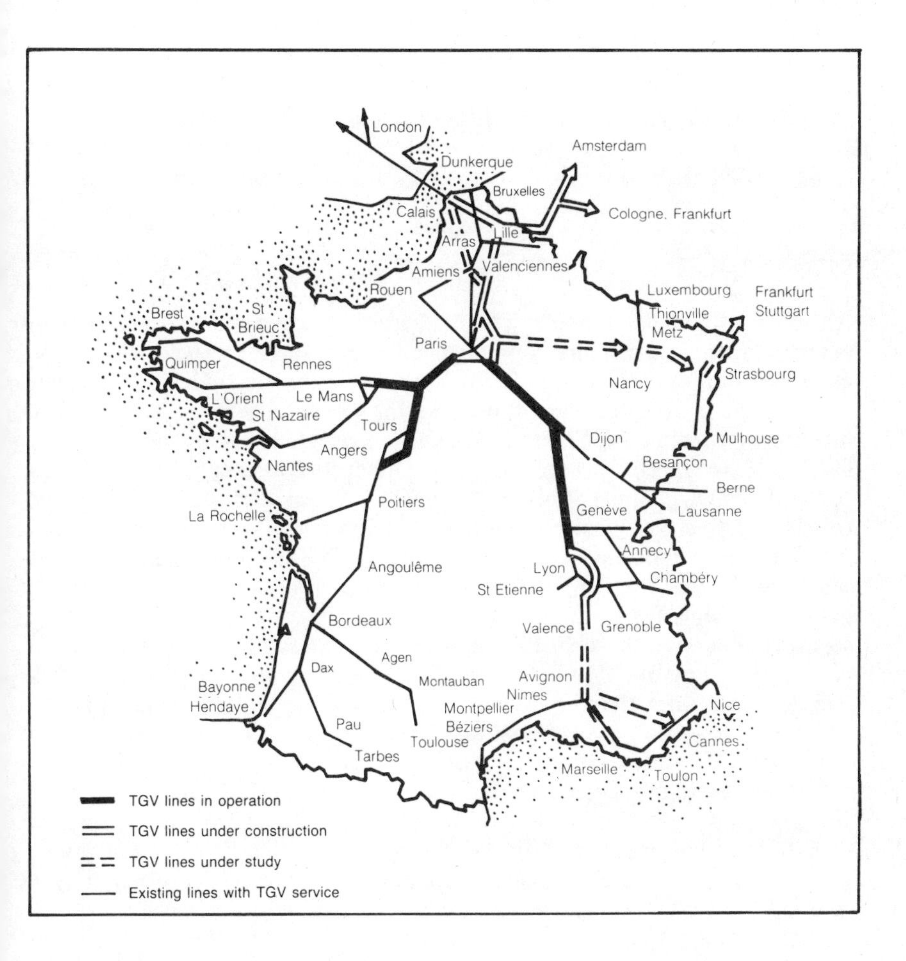

Figure 7.5 *French TGV network*

effect. This involves little compensatory expenditure — a *centre de la mer* in Boulogne, for example, pre-dates the Tunnel project, and was written into the first *contrat de plan Etat-région* — but is chiefly a strategic investment in tourism infrastructure, such as upgraded hotels, English-style bed and breakfast accommodation, restaurant facilities, and so on.

Beyond Nord-Pas de Calais, French planners are using the gradual emergence of a national TGV network, shown in Figure 7.5, to promote France internationally, and to focus development on depressed parts of the country. To take the single example of the East of Paris, routeing of the TGV Interconnection (a kind of *périphérique* for trains) to the east of the capital has allowed the TGV to be linked to both Roissy and Orly airports, and has permitted substantial development at Marne-la-

Vallée, where EuroDisneyland is being built (with its own TGV stop), and where Universal Studios is similarly thought to be considering locating its European theme park (if it does not locate at Rainham Marshes, to the east of London). On the basis of an innovative transport strategy, the east of Paris is thus undergoing major redevelopment, and is being hailed as a model example of urban regeneration.

Constrained market solutions in Britain

Although the official policy of the British government is to allow the market to decide as many aspects of Channel Tunnel policy as possible, in the realm of regional economic development the market is heavily constrained in Britain. This is because the areas of greatest development interest as a result of Channel Tunnel construction (many of which are in Kent) are carefully controlled by Green Belt, Area of Outstanding Natural Beauty, Site of Special Scientific Interest and other restrictions, which effectively reduce the number of developable sites to a mere handful. Research undertaken by Kent County Council (KCC) in mid-1989 suggested that at that time there were perhaps 1000 ha of land with planning permission for industrial use in Kent, compared with at least 30 or 40 times that area in Nord-Pas de Calais (which is bigger than Kent, but only by a factor of 3.5). Indeed, such is the availability of land in Nord-Pas de Calais that the very notion of developable land becomes difficult to determine: it is scarcely an exaggeration to say that the entire region is open for business (Roberts, 1990).

In these circumstances, the roles of public authorities and private developers are drastically altered. In areas of high private-sector interest (and low development land provision), it is not necessary for the public sector to pump-prime. Instead, it is able to restrict itself to merely a controlling function, allowing development which it considers to be desirable, and forbidding all other investment. In part, this is the current situation in Kent. There are, however, a number of ways in which Kent departs from this description, and there are many ways in which the rest of Britain fails to conform to it. Thus, British public authorities are not restricted to a controlling function.

In analysing regional policy responses in Britain, therefore, it is necessary to pay careful attention to economic context. It is difficult to criticise the stance taken by planners in affluent and highly desirable districts of Kent, such as Ashford and all points west (Tonbridge, Tunbridge Wells, Sevenoaks), who select high-quality projects from a flood of development interest. What is debatable is the orientation of planners at more strategic levels who, faced with low development interest in other parts of the county and country, (ought to) have the responsibility of channelling development to them. This function is currently being neglected in the UK.

An early proposal, floated by the Kent Impact Study Team, to institute an East Kent Development Agency to manage development in affluent Ashford and Shepway, in Canterbury, and in struggling Dover and Thanet, was undermined by the successive withdrawals of Shepway and Ashford, the twin linchpins of the scheme.[4] Having no executive power of its own, the JCC was unable to step in to rescue the proposal. At county level, KCC acts both through its own Economic Development Department, and through the semi-autonomous Kent Economic Development Board to promote the county as a prime business location, but little direction of investment towards the county blackspots is undertaken by these bodies. Rather, the county as a whole is marketed in an international arena. KCC has recently started to advertise the county in partnership with the Nord-Pas de Calais Regional Council, under the auspices of the new Euroregion which the two form, but this activity, though clearly valuable, is essentially promotional and not directive.

A similar pattern is repeated throughout the British system. At regional level, SERPLAN, the South East Regional Planning Conference, issues advice, but has no executive authority. At pan-regional level, bodies such as the North of England Regional Consortium express the views of large groups of local authorities, but have little impact on government thinking, which is determinedly wedded to the notion of market outcomes. The most visible example of this policy, and the clearest contrast with French public policy, is treatment of the Channel Tunnel rail link in the United Kingdom. The transport implications of this treatment have already been discussed. What has not yet been considered are the regional economic consequences of the policy decision, first, to view the British rail link issue as essentially a pure transport matter, and second, to see that transport matter as chiefly a problem of getting passengers across Kent from Cheriton to London. Each decision runs counter to French practice, and may be contrasted with it.

Thus, where the French have taken the TGV first to, and then through, Lille on grounds of *aménagement du territoire*, British policy-makers in the Department of Transport have so far refused to plan the regional economic consequences of the Channel Tunnel either between London and the Channel, or beyond London. It is partly on these grounds that rival promoters of a high-speed Kent rail link have criticised BR's scheme. Both Ove Arup and Rail Europe have sketched out a route which enters London via the east — the so-called 'right hook' — with a major international passenger and freight terminal at Stratford, east London. On this basis, they not only allow for substantial development at Stratford, in conformity with the wishes of the London Borough of Newham (Colin Buchanan and Partners, 1989), but also raise the possibility of development further down the line on derelict or near-derelict sites: hence plans for a Rainham Universal City station on the Arup scheme, to attract Universal Studios, and for stations to link with both the M25 and the major retail park at Blue Water Park, Dartford. The parallel with French practice, both in Lille and Marne-la-Vallée (also a possible location for Universal Studios), is clear (*The Economist*, 1989).

Arup and Rail Europe also claim that their schemes provide better links to the North and West of England than does the BR project. However, this claim needs to be treated with some scepticism. On the passenger side, it is clear that, without a dedicated link between Stratford and King's Cross, the service offered to international travellers is markedly inferior to that planned by BR. Quite simply, Stratford is not central London, and its links to the rest of the country are not as good as are those of King's Cross: both passengers for London and through travellers are, therefore, at a disadvantage. It should also be noted that the extra £1 billion that such a link would cost is not usually included in Arup and Rail Europe costings: they apparently hope to factor this particular expense out to the public sector. Once this extra cost is included, the difference in price between BR and rival schemes is minimal. On the freight side, the possibility of a terminal at Stratford is opened up by the rival schemes, but in rail terms this is little different to BR's plans for a marshalling yard at Wembley, and in road terms it offers the unappealing prospect of a distribution depot close to the heart of London. Where the rival schemes do hold an advantage over BR is in planning four-lane dedicated tracks across much of Kent, which would allow for freight movements on this track rather than, as in BR's plans, on the residual capacity on existing lines in Kent which would be released by a dedicated passenger link.

In these circumstances, the problems facing the North (including Scotland), the West (in particular, Wales) and the South-West (not to mention both Northern Ireland and the Irish Republic) are grave. As will be noted in Chapter 10, economic development in the wake of transport investment is by no means assured; but economic underdevelopment in its absence is well documented. The prospect which faces British regions beyond London is increased peripheralisation in post-1992 Europe.[5]

7.3 Conclusion

The paradoxical possible outcome of French *aménagement du territoire* and British policy mess, concerning both the Channel Tunnel rail link and subsidiary questions, is that the market will actually be a strong determinant of economic development in France, and that the public sector will play an important role in Britain. This is because the possible landtake on the French side is vast, and because intercommunal rivalry remains a strong constraint on strategic action. It is, for example, likely that international conference centres will be built both at the Tunnel mouth, in the ZAC, and just down the road in Calais. It is by no means certain that both will be profitable. In Britain, by contrast, the possible landtake in the most developable Tunnel-related sites — mainly in Kent — is small, thereby allowing local authorities a large say in the types of development that are permitted. At the same time the early enthusiasm of British investors for projects in Nord-Pas de Calais (75 per cent of total planned property investment in the Calais region was British in origin, according to Chaplain (1990)) has waned as a result of both the recession and the recognition that Europe only begins at Calais.

However, this paradox is merely a reflection of the different economic contexts in which Tunnel policy-makers operate in Britain and France. It does not present a judgment on their revealed abilities. That judgment must be that, despite the problems which local rivalries create in France, strategic rationality has been a great deal more developed there than it has been in Britain. Even in the long term, in which a fairly broad consensus around investment in infrastructure links exists, British policy-makers have been unable to develop a satisfactory policy. Instead, constrained by the ideology of a government which refused to sanction public investment in a dedicated rail link to the Tunnel, they presided over the rail-link fiasco which stands in such stark contrast to its French equivalent.

Notes

1. See Chapter 6, pp.102–3.
2. See p.136.
3. The argument which was subsequently deployed by BR — when, contrary to assurances given during passage of the Channel Tunnel Bill, it sought a mere two years later to build a dedicated line from Cheriton to central London — was that it had witnessed declining passenger volumes totalling some 15–20 per cent between 1970 and 1984, and an unexpectedly dramatic upturn in rail usage at a rate of 5 per cent per annum after 1985. Thus, a Channel Tunnel rail link which was said to be unnecessary in 1986, was held to be vital in 1988.
4. See Chapter 6.
5. See Chapter 10 for a full discussion of the regional development implications of the Tunnel and associated infrastructure investment.

8 *Public policy and the fixed link*

Having set fact against theory, it is time to draw some conclusions from the Channel Tunnel experience about the conduct of public policy in Britain and France. In this chapter, this exercise will be undertaken from the perspective of political science: what does the Tunnel reveal about the British and French political systems? In the next, a different perspective will be adopted, and the question which will be asked is, what does the Tunnel reveal about mechanisms for provision of transport infrastructure? There is some overlap between the two questions, but for the purposes of analysis they may be distinguished.

8.1 Policy networks, communities and styles

An extensive literature testifies to the vast complexity of policy networks, communities and styles in both Britain and France. Running through this literature is a clear distinction between the British and French policy styles. As Richardson *et al.* (1982) argue in introducing a collection of essays on policy style, the British have a predilection for consultation and for bureaucratic accommodation (see also Jordan and Richardson, 1987), whereas the French oscillate between stagnation and radical policy change. The interest of the Channel Tunnel experience is that it qualifies the standard assessment of the British policy style, and elucidates some of the conditions of operation of the French policy style.

A key feature of the Tunnel is that it has both national and regional/local impacts. It is an instrument of national policy with significant repercussions at the subnational level. At some stage, then, subnational government is implicated in its planning. The key difference between the UK and France that is revealed by the Tunnel experience is the stage at which, and conditions under which, this implication is unavoidable. Put crudely, subnational government in the UK was brought — or allowed — into the policy network approximately three years after its French counterparts had been admitted to the French network. This suggests that even in Britain, where policy-making is supposedly premissed on consultation, central government is (not surprisingly) keen to exclude potentially awkward actors from the policy process for as long as possible. The means by which it was able to do this in the Tunnel case require investigation of the nature of the policy networks in which central and local government are represented in Britain and France.

In Britain, policy networks in which local government is implicated have often tended to develop by one of two routes. On the one hand, central government departments, such as the Departments of the Environment, of Education and Science, and of Transport, have established direct links with local government departments, sometimes using regional offices to establish close local ties. On the other, policy networks have developed through national associations of local government officers and members. In opposition to the traditional view of the British system as strongly decentralised, an alternative perspective has recently emerged which stresses the significant amount of centralisation generated both by direct central–local links, and by the large number of national associations which operate at subnational level. These associations are both professional and political, and, being creatures of the local government world, represent in some sense self-centralisation of a decentralised local government system. A series of needs and prejudices may be cited to account for this phenomenon: the need for information from the centre, and for access to it; professional and political prejudices which operate throughout a government system; and even a pervasive British executive prejudice, held by people who operate in the local government world. This, then, is Rhodes's (1986) national world of British local government.

What is notable about it is that its undoubted centralisation is to a large extent uninstitutionalised. The associations through which policy networks are constructed to link central and local government departments and politicians are in many senses *ad hoc*. They have no constitutional significance, and no real legal basis. Indeed, they tend to be in a vulnerable position in the courts, and suffer frequent defeat. This may be explained by the fundamental British notion of Parliamentary sovereignty, which is reinforced by executive prejudice — on the part of the judiciary — and by the strength of the British common law tradition, which is characterised by a developed individualism, and which has little understanding of associations in consequence. Thus, in Britain, intergovernmental policy networks have been constructed on the basis of mutual convenience, which, in the absence of formal institutionalisation, becomes the condition of their existence.

In France, the situation is very different. Again, extensive policy networks link centre to periphery, but they are constituted in distinctive ways from their British counterparts. At an official or professional level, they depend on the prefectoral system, and on the extensive field services of the French state, which give French central government a presence in the localities which cannot be matched by the British state (which possesses neither of these attributes). At a political level, they depend to a large extent on the system of *cumul des mandats*, by which dual representation of centre and periphery is effected by such people as Mauroy and Delebarre.

Centralisation, and, in recent years, decentralisation, are thus heavily institutionalised in France, and are indeed codified to a significant degree. The nature of this institutionalisation is subject to change by, for example, the series of decentralisation and deconcentration measures instituted, or ratified (Mény, 1987), by Mauroy and Defferre at the beginning of the 1980s, which fundamentally altered many of the operations of the French governmental system. The fact of it is not, however, alterable. Thus, although in some respects decentralisation has removed the prefect from his position as ultimate political fixer, and fulcrum of many policy networks, it has not deinstitutionalised intergovernmental relations in France. In small local authorities, such as rural communes and *départements*, the prefect and the field services of the state remain crucial to management of the locality. In large local authorities, such as major urban centres, many *départements*, and all regions, local *notables*, holding a central and a local mandate, retain control. In each case, links between centre and periphery are firmly institutionalised.

The extent of this institutionalisation must also be stressed. Institutionalised French policy networks are not just intergovernmental, but extend to encompass social and business elites through such mechanisms as *comités interprofessionnels sociaux et économiques* (CISE), which must be consulted by, say, regional councils engaged in major policy debate. Such enforced consultation does not necessarily give the CISE leverage over decision-makers at any level of government, but it does give them priviliged access to the decision-making process.

The result of this basic difference in policy networks — often voluntarist in the UK, institutionalised in France — is that capacities for state action are actually the reverse of what have traditionally been held to be the preferred policy styles of the two states. The British state has the ability to operate outside the constraints of local politics, simply by declining to consult through traditional mechanisms. Crucially, the periphery has very little power to prevent this, as is clearly shown by the Channel Tunnel experience. The French state has no equivalent ability, and is forced at least to hear local views on all major decisions. Institutionalisation of policy networks offers the periphery guaranteed access to the centre of French government. Thus, French policy-makers have only a constrained capacity for heroic action, whereas their British counterparts face little restriction should they choose to act outside their customary incrementalism. A dual policy style is thus more likely in Britain than in France. This is certainly what has been witnessed in the case of the Channel Tunnel.

8.2 Formulation and implementation of the Channel Tunnel project

It is difficult to characterise as heroic any policy which takes more than a century to implement. In some respects, however, the Channel fixed-link project, first seriously mooted in 1802, briefly implemented in the 1870s and 1880s, and finally to be completed in 1993, fits this characterisation. It does so in different ways in Britain and France. Hayward (1974, p.399) has identified a dual sense to heroism in policy-making: 'both an ambitious political exercise in rational decision-making and an ambitious assertion of political will by government leaders'. On the British side, fixed-link policy-making has been characterised by an assertion of political will, but by little rational planning. On the French side, little political will was exerted, but rational planning has been sought.

This distribution of political will might be thought strange, both because the British bear almost sole responsibility for decades of (far from heroic) obstruction of the fixed-link project, and because one of the prime features of the French political system is often thought to be its authoritarian cast. However, it is in part the very history of British blocking and delay which created the climate in the 1980s for heroic resolution of a very old problem. And, by the same token, it is the length of French commitment to the project which has generated often unheroic fixed-link policy-making in recent years. This is an important contextual point, for it suggests that it is very nearly impossible for a heroic exercise of political will to be undertaken in conditions of near-total agreement about policy priorities. The converse is also true. The distribution of rational planning will not be thought at all strange: it is what has long been said about the two systems.

What this demonstrates is the care which needs to be exercised in handling the concept of heroic policy-making. At first glance, French fixed-link policy-making looks heroic in the dual sense intended by Hayward. Not only was authorisation for the Channel Tunnel itself secured with minimum fuss — unanimously in the National Assembly — but also road and rail links to it were agreed rapidly and in apparent conformity to a rational plan. Yet this apparent assertion of political will on behalf of a team of rational planners is in reality nothing of the sort. On the one hand, virtually no political will had to be exerted, because outright opposition to the Tunnel project itself, and to the proposal to build infrastructure links to it, was minimal. On the other hand, many groups and interests had their price for acquiescence in the scheme or in links to it, and rational planning was sometimes compromised by accommodation to local or regional interests. This is certainly not to argue that there was nothing rational about the planning of French infrastructure links to the Channel Tunnel, simply that the planning process in France was more humdrum than is sometimes thought.

By contrast, there certainly was an element of heroism to British fixed-link policy-making. A crucial exercise of political will was required definitively to launch a project which had been, and which clearly remained, controversial, and almost emotive in some parts of the country (notably Kent), and among some sections of the population (which did not only comprise retired colonels). In this sense, the British government, and in particular Mrs Thatcher, undoubtedly acted heroically. Hayward's 'ambitious political exercise in rational decision-making' was, however, absent on the British side, because state planning was felt by the government to be an irrational means of proceeding, being clearly inferior to market mechanisms. At this point, the British government was not heroic in Hayward's sense.

Looking, then, at the two components to heroism, British fixed-link policy-makers were long on political will, but short on concerted decision-making, while their French counterparts were short on political will, but relatively long on rational decision-making. What this distribution seems to suggest is that Hayward's dual elements of a heroic policy style are actually incompatible. Policy may be either wilful and non-rational, or consensual and rational, but it cannot be wilful and rational because the very concept of political will contains within itself a refusal to consult with interests which a modern polity must bring on board if policy-making is to be successful.

8.3 The conduct of public policy in Britain and France

What this analysis presents is, then, a verification of Ashford's (1982) thesis. It is the British who can get away with being dogmatic, because policy-makers in central government are not tied into policy networks in the manner of their French counterparts, and the French who are forced to be pragmatic, by the political and administrative webs in which policy-makers at all levels are necessarily implicated.

This analysis also confirms the conclusion which Ashford draws from these different policy orientations, and from the different distributions of will and concertation which have been revealed by the Channel Tunnel experience: that the result is often policy mess in the UK, and a degree of policy success in France. The point here is that British governments have a capacity for dramatic action — as in the fundamental restructuring of the governmental system during the 1980s — which is almost unmatched in other European states. This capacity was frequently exploited by the Thatcher government, not only in its spectacular abolition of the GLC and of the six metropolitan county councils, but also in more gradualist restructurings of a number of policy communities, through trade union and local government reform, and also through policy initiatives in the industrial sphere and in many areas of the Welfare State. By these means, successive Thatcher governments diminished the power of many intermediate institutions, and made some progress towards the Thatcherite ideal of a strong individualistic society regulated by a strong centralised state, an ideal which is fully explored by Gamble (1988).

However, this capacity for dramatic action, for both the heroic gesture and the more measured reconstruction of policy communities throughout the operational areas of the British state, is a capacity for successful policy-making only in areas where intermediate institutions can be bypassed with impunity. Vickers and Wright (1988) identify the British privatisation programme as in many ways just such an area of policy-making, arguing that the capacity of Thatcher governments to restructure policy communities was important in removing obstacles to privatisation. Specifically, they argue that trade unions have been successfully marginalised, thereby smoothing a number of paths to privatisation, and that even management has on occasion been excluded from decisions to privatise. 'The announcement of the sale of BT', they write, 'apparently came as a complete surprise to management, but after they had displayed some enthusiasm for privatisation they were brought back into the policy community' (ibid., p.16). The Channel Tunnel represents the alternative scenario, in which an attempt to exclude intermediate interests — such as, in particular, local government — led to policy failure, not policy success, because in the final analysis a substantial degree of co-ordination and *concertation* was necessary to successful policy formulation and implementation.

The contrasting experience of the French government, which was locked into policy networks which it could neither restructure nor bypass at will, and which was, therefore, obliged to engage in mechanisms of *concertation*, demonstrates that some degree of policy success was thereby almost institutionally assured. French policy-making is not insured against mess, but it is at least prevented from falling into some of the traps which can befall the British. To British eyes, the French system may look horribly conspiratorial, and in some senses it is, but the obverse of this is a degree of co-ordination which British policy-making can all too easily fail to attain.

What is particularly interesting about France in comparison with Britain is that both countries are characterised by tightly knit policy communities (in contrast to a number of other European states). In France, as in Britain, these policy communities tend to be dominated by technocrats, and by a small group of politicians (the balance between the two groups perhaps being different in the two countries). Yet policy-makers at the centre of French government do not have the freedom of manoeuvre that is sometimes allowed their British counterparts because of the high degree of institutionalisation of French policy networks.

The result is that the conduct of public policy in Britain can be — though, of course, is not always — erratic. In France, this possibility is greatly reduced. As has been seen, in the case of the Channel Tunnel this potential for erraticism was to a large extent realised by successive Thatcher governments, with predictably unfortunate results.

Part Three

Regional development and European integration

9 *The transport policy debate*

A major theme of this book, introduced in Chapter 2, is the question of whether the present Channel Tunnel project may in any sense be termed a great planning disaster. Technically, with the increasing likelihood that a Tunnel will be operative by the mid-1990s, the project could only qualify as a positive planning disaster. However, a sense of negative disaster also hangs over what many see as the British government's failure to integrate the Tunnel system into a coherent transport network. This consideration raises a series of larger questions which are now being addressed by transport analysts in Britain and Europe, and must be set in this broader context if the full consequences of the decision to concede Channel Tunnel construction and operation to the private sector are to be drawn.

This chapter considers, first, the efficiency implications of the decision to 'privatise' the Tunnel; second, the argument used by the British government that the market is the best judge of transport need, and the best means of meeting that need; and third, alternative (non-market) means of supplying transport in the context of the development of moves towards a European transport policy.

9.1 The Channel Tunnel as an efficiency test of privatised transport infrastructure

The Channel Tunnel can be seen as the flagship of the Thatcher government's transport 'privatisation' strategy, and forms a good context in which to measure the British approach to transport provision against alternative continental approaches. In essence, this can be managed in a dual perspective, looking first at the 'internal' impact of privatisation on a discrete piece of transport infrastructure (in this case the Tunnel), and second at its 'external' impact on a complete transport system.

Internal consequences of Tunnel privatisation

Peter Hall's (1980) definition of a positive planning disaster is a scheme which many informed people recognise to be a mistake, but which nevertheless continues to be pursued (and financed). Does, therefore, conceding decision-making to the private sector, where it has to compete for funds with a wider range of other projects, mean that such a mistake cannot be made? It is not clear that it does. In one sense it could be argued that a principal reason for the rehabilitation of successive Channel Tunnel projects was pressure from private-sector interests: transport users wanting improved cross-Channel facilities, and construction companies and financiers looking for a profitable major project. However, a project like the Channel Tunnel requires government interest and support, and once it has received such support there is an inexorable logic pushing towards its completion, with or without guarantees, provided certain critical early hurdles are cleared. Hence the Bank of England's continued monitoring of progress, which seems closer than is required to safeguard the interests of British financial institutions implicated in the project.

Privatisation, in the broad sense, contributes to a number of political objectives, such as reducing public spending, controlling trade unions, and widening share ownership, but its main aim and justification is — or should be — the promotion of market efficiency (Vickers and Yarrow, 1988). In conditions of competitive markets, private enterprise is shown by standard economic theory to be an efficient producer and distributor of goods, and privatisation of firms in markets approaching perfect competition has usually been unproblematical. In conditions of near- or actual monopoly, genuinely private enterprise is less certain to be efficient in either production or distribution, and privatisation of quasi-monopolies has required simultaneous establishment of regulatory agencies with the specific brief of promoting efficiency.

The efficiency implications of conceding the Channel Tunnel to private enterprise reflect a wider concept of privatisation. The granting of a concession implies the granting of a monopoly through a process of competition: selection of an agreed service at the lowest tendered price introduces the virtues of competition into naturally monopolistic markets. The consumer benefits, not directly through competition in service provision, but indirectly through competition for the right to provide the service. Such right is granted for a limited time only, with no guarantee of renewal, and subject to compliance with certain regulatory controls. It is unproblematical and generally successful in areas in which an agreed service is easily specified. It becomes increasingly problematical, and is less certain to promote industrial efficiency, as difficulties in service definition mount. The need for regulation to promote efficiency rises directly with such difficulties.

The Channel fixed-link project is at the most difficult end of the continuum of service definition: no more than broad parameters were laid down by the two governments in the April 1985 *Invitation to Promoters*, provoking a variety of fixed-link proposals. In these circumstances, the present Tunnel project is constituted as a privatised and regulated quasi-monopoly. It is similar in some respects to other recently privatised industries, such as telecommunications, air transport and a number of utilities, except that it does face competition from other types of transport, and the governments did not attempt to introduce the form of price regulation typical of privatised utilities. Regulation is of minimum service levels at the busiest and slackest hours of the year, of safety, and of other operational conditions. Otherwise there is no attempt to restrict Eurotunnel's commercial freedom.

The advantages of greater efficiency that would result from concession of the Tunnel to the private sector may be reviewed in terms of the two phases of the Tunnel project itself, construction and operation. In the construction phase, the private sector was seen as being more likely to control costs and construction progress. The Tunnel would thus be built more quickly and cheaply, thereby generating considerable social benefits. At this stage the government's regulatory interest is concerned mainly with detailed aspects of project design, and thus with the type of service eventually to be offered. However, regulatory agencies face a dual information deficit: they can control neither the completeness nor the timing of information disclosures by Eurotunnel and its contractors, TML. In these circumstances, they are likely to encounter clear monitoring difficulties. Lack of complete information is an evident impediment to effective regulation. Late receipt of information can present regulators with an invidious choice: assent to the proposals, or jeopardise the project's commercial viability.

In practice the pressure of time has meant that the Intergovernmental Commission (IGC) has received drafts of the *avant projets* it is required to approve during construction. This leads potentially to a series of piecemeal approvals without viewing the whole, and makes ultimate rejection much more difficult. The separation of passengers and vehicles on shuttle trains, for example, which has provoked extensive debate in the British press, is one question which could be seen to be determined more by reference to Eurotunnel's profit motive than by safety criteria.

In the operation phase, when Eurotunnel will be licensed to provide the service agreed in the construction phase, the focus of regulation shifts to control of Eurotunnel's quasi-monopoly market power. The Tunnel will be in an unusual economic position, as described in Chapter 3: with high fixed and low variable costs, its structure is very different to that of existing operators, and puts it in the position of potentially being able to undercut cross-Channel prices until competitors are forced out of the market. Whether Eurotunnel would be allowed to pursue such a strategy is unclear from the concession agreement, where regulatory powers are at their most vague: Clause 12 states that Eurotunnel is 'free to determine [its] tariffs and commercial policy', while remaining subject to 'national or Community laws concerning competition or the abuse of a dominant position'. It seems likely that the threat of a referral to the Monopolies and Mergers Commission (MMC) (or its EC equivalent) would be enough to dissuade Eurotunnel from exploiting its position too ruthlessly, though at this distance such a judgement is speculative.

Clearly, however, the MMC took the view in December 1989, when ruling against the proposal by P&O and Sealink to co-ordinate their services, that lack of competition between the ferry companies would lead to higher prices, and that Eurotunnel could fix its prices by reference to these higher prices (Monopolies and Mergers Commission, 1989b). The MMC's understanding was essentially of Eurotunnel as an accommodating oligopolist, rather than as a ruthless competitor insistent on achieving an eventual monopoly position.

In either outcome, however, judged by the narrow criterion of market efficiency, the Tunnel scheme thus displays the classic regulatory problem of transport: the instability between non-competitive oligopoly and destructive competition leading to monopoly. This is heightened here by differences in modal characteristics and cost structures. The information deficit of the regulators makes it difficult for them to fulfil their duties in the specified areas of safety, security, the environment and national defence. The extreme vagueness of the operating contract itself places a large responsibility on political will in the operating phase.

Thus, while a large amount of regulatory machinery has been built around the Tunnel, it is likely that real economic control of the project has been transferred to the banking syndicate, which becomes a more effective monitor of the project than are public authorities. It is the banks which control finance at each stage, by monitoring construction progress before allowing Eurotunnel to draw on its agreed lines of finance. It is this debt which has the first call on Eurotunnel's operating surplus and which will, therefore, dominate commercial policy considerations in the early years of operation. In these circumstances, market efficiency may have been sacrificed to monopoly practices by privatising the project.

Efficiency must, however, be judged more widely. To begin with, market efficiency itself can be read in a wider context. It is possible that ultimately Eurotunnel will, despite its initial weakness, be a more efficient manager of TML than public authorities would have been, and that the Tunnel will be built more cheaply as a result: some research suggests that contractor–client relationships are better handled by the private sector, and that a clearer project basis is likely to be established (Morris and Hough, 1986). It is also possible that concession of the fixed link has resulted in a better project: a central belief of the Thatcher government was that competition would foster a process of discovery which would generate more efficient fixed-link schemes than could be devised by government departments. The fact that the discovery method has produced a scheme which closely resembles the public-sector 1970s project suggests, however, that this efficiency gain should not be exaggerated. How far this is an indicator of the efficiency of the public sector in the 1960s and 1970s, and how far an indication that the ultimate selection was undertaken by public-sector officials is an open question. Their management of politicians desiring more glamorous drive-through schemes is an interesting issue (Henderson, 1987).

In a wider context still, the efficiency of the present Tunnel project may be assessed in terms of policies which have been developed around it. This aspect of privatisation — the impact of privatised industry on the residual public sector and public policy — is sometimes neglected. It is, however, central to discussion of the Tunnel, which is only one (small) part of a substantial infrastructure investment programme currently being developed and undertaken in Europe. The question that is addressed in the next subsection is how privatisation of the Tunnel has affected wider public management of the new transport system which it sets in place.

External consequences of privatisation

It is important when discussing policy responses to the Tunnel not to treat the public sector monolithically: policies have been developed by public institutions at many levels, local, national and international. However, one important division exists, between the two national political systems within which policy responses are articulated. Despite the emergent European dimension to transport policy, at the present time it is still the national systems which remain primary determinants of public policy. The distinction between them gives a dual character to this subsection, and provides two distinct assessments of the impact of privatisation on public policy.

Economic and ideological reasons may be given for different public responses to the Tunnel in Britain and France, but the more important reasons are institutional. It is true that the economic coherence of the project is evident at both local and national level in France, in contrast to conflicting local and national perspectives in Britain, and that this has facilitated policy responses. It is also true that the pragmatic desire of both left- and right-wing French governments to seek workable solutions to problems posed by the Tunnel contrasts with the ideological insistence on pure private enterprise of the Thatcher government in Britain, and that this, too, has eased policy co-ordination. However, as the following account shows, institutional differences have had a more substantial impact on policy responses in the two countries.

A key issue in both Britain and France has been road and rail links to the Tunnel. In each case, British policy responses have been more fragmented than French. The main explanation of divergent responses in the two countries is institutional. Among both politicians and administrators, French policy networks extend from the centre to the periphery in largely unbroken fashion, with the result that a degree of co-ordination of national and local policies is attained in even the most unpropitious circumstances. In Britain, by contrast, the lack of such networks, or a denial of their existence by key figures at the centre of government, means that it is possible for national and local policies to be in fundamental conflict. The decision to build a TGV line from the Tunnel to Paris was clearly part of a supporting policy for the Tunnel scheme (it was announced in October 1987 just before the main public share flotation by Eurotunnel). Its economic logic was enhanced by its being part of the proposed northern Europe TGV network announced at the same time. There was a clear desire to make Paris the central focus of this emergent high-speed rail network.

However, it is hard to believe that, had the French been faced with the rather different geographical situation and consequent problems which have arisen in Britain, they would have acted as British policy makers have (re)acted. Institutional networks, with their developed links between centre and periphery, would have prevented the fragmented and often reactive responses that have been, and continue to be, witnessed in Britain. The French system, seen at its most developed in the person of Pierre Mauroy, was critical in securing the finance necessary to persuade SNCF to route the TGV line via the centre of Lille in order to give substance to local development plans. In these circumstances, a co-ordinated strategic response is more likely to emerge, if only because there will always be someone at the centre who is not allowed to ignore the problems and desired solutions of the localities.

Beyond personalities, the structure of policies in France, with the concepts of *utilité publique* and of *projets 'grands chantiers'* being applied to the Tunnel despite its private-sector position, enables both co-ordination of supporting policies and direct access to Prime Minister and President to avoid any interdepartmental wrangles which can beset projects with transport, physical planning, regional development and environmental consequences.

It is, then, French institutions which have ensured that the Tunnel is seen as part of an integrated transport network. Strategic vision of this kind has conspicuously not been developed by British institutions.

First, there was no easy focus of regional development issues such that an acceptable coincidence of national and regional interests could emerge, as we shall discuss at greater length in Chapter 10. The British government stuck rigidly to the view that, since it was difficult to identify the direct impact of infrastructure on regional development, there should be no specific policy to develop this aspect, other than relying on usual monitoring. Second, transport had slipped badly down the list of the government's priorities in the 1980s. Public transport was seen as inefficient due to excessive regulation and public subsidy. Road schemes were coming under increasing environmental scrutiny as what were seen as the final sections of the key motorway network were completed. Moreover, the general philosophy relating to congestion and regional imbalance was that the market would respond; rising congestion and house prices in south-east England would curb excessive development; and increased competitiveness in other regions would derive from reductions in excessive local council spending on subsidies.

British Rail lay at the centre of this debate, despite being one of the least subsidised railways in Europe. Its division into separate sectors, each with its own financial objective, the removal of all subsidy from two of these sectors (Railfreight and Inter City), and the severe reduction of public service obligation grants to Network South East and Provincial (now Regional Railways), brought about much-needed increases in efficiency. This process was assisted by referral of BR's monopoly position to the Monopolies and Mergers Commission (1988; 1989a). On investment, commercial criteria determined by the Treasury dominated, although BR could be grant-aided where there was a determinable social benefit, such as the relief of road congestion. It was envisaged that this would only occur in highly congested urban situations. The likelihood of major investment in new lines was not foreseen. BR had after all pioneered high-speed regular-interval passenger services; the costs of constructing new infrastructure to raise speeds above 200–225 km/h were seen as prohibitive. Above all, it was to be users who should pay for any improvements, as they were seen to be the beneficiaries — any social benefits would be captured through the revenue generated by increased travel. The crowning, though economically curious, argument was that since rail carried only a small fraction of total passenger-kilometres anyway it was wrong for it to be heavily subsidised for new investment to benefit the few. This argument ignored the extent to which the much-vaunted level playing field of competition had been achieved.

When it came to the question of a rail link to the Tunnel, a combination of arguments was used: BR was being allowed to invest to the order of £1 billion anyway in upgraded track, the Waterloo terminus and new trains; users would benefit and so users should pay the full costs (including all environmental costs); the government did not subsidise ferries and airlines on international routes so it should not subsidise rail (and anyway section 42 of the Channel Tunnel Act 1987 explicitly prohibited such subsidy). Each of these arguments raises problems.

Had BR been allowed to consider all the alternatives for investment, some of the new investment might have been unnecessary if earlier development of a new route had been considered. Road traffic does not pay its full marginal social costs of specific environmental damage; it may pay globally, but the individual user of a stretch of road pays no more than the non-user. Rail is in a different position to other international modes since with the Tunnel it is the only one offering through-transport free of mode changes. Airlines and ferries depend on often subsidised access modes (urban public transport and congested roads), and both are allowed to cross-subsidise between different services in a way which BR's international services would find very difficult.

As the argument continued on these issues, and the British government came under increasing criticism with each further development of high-speed rail in Europe, it reverted to attacking the notion of high-speed rail itself. From arguing (erroneously) that the French TGV was excessively subsidised, criticism was redirected to its lack of energy efficiency and levels of noise pollution. Here, the comparison was with British high-speed trains, which at 200 km/h are more energy efficient per seat-kilometre and quieter; but such a comparison is only relevant if higher-speed trains do not divert traffic from road and air. To the extent that they do, real comparison should be with these modes.

Throughout in Britain, the concern seems to have been to avoid any semblance of a coherent strategy, regardless of whether the ultimate outcome was right in terms of economic efficiency. This suggested that the government was always being reactive rather than proactive. It was left to local government to make the running in terms of pushing BR to decisions over services, freight depots, and so on. Rather as with regulatory control of the Tunnel itself, the British government passed the initiative to others, even when they had no clear direction of purpose. When rival rail-link schemes were proposed, there was no obvious way they could be compared with BR's own scheme, and ultimately the government left the job of comparison to BR, as the body statutorily charged with running railways.

9.2 The market and transport infrastructure provision

The argument for public provision of transport infrastructure rests on four main assumptions: first, that there are scale economies and network effects which only the public sector can exploit; second, that the time-scale for planning and developing infrastructure requires substantial finance which the public sector can provide most cheaply; third, that there are externalities which affect non-users and which would not be taken into account in determining optimal scale; and fourth, that there is a tendency to monopoly which would require regulatory control.

However, against this can be ranged arguments in favour of private-sector involvement. The scale economies argument may have been over-played recently, and network economies may also be viewed as an excuse for inefficient cross-subsidy. There may be an argument for co-ordination in marketing complementary infrastructure, common ticketing, and so on, but this does not necessarily require a single monolithic transport organisation.

While it is true that typically the public sector can borrow more cheaply, if the private sector can build more efficiently at lower cost, the overall resource cost can be lower. The public sector has often been an inefficient manager of large projects. Because of its position it has often been seen as not having a limit to its financial position in the way a private-sector manager would, which makes it vulnerable to exploitation.

The externalities question varies with type of infrastructure. The critical issue is how far the benefits of the system can be captured by a private-sector operator. This depends largely on its freedom from direct competition, as this will determine the present value of the concession itself. As Alastair Morton (1989) has argued, the Channel Tunnel's uniqueness is its value, which makes private finance viable in a way which could never be true of a rail link. This minimising of competitive risk increases the rent inherent in a project (Gerardin, 1990), but raises the problem of monopoly. In such a context, although a privately financed infrastructure might be viable, construction needs to be delayed to increase rental value, whereas a social valuation might indicate earlier construction (Kay *et al.*, 1989). Nevertheless, in situations of scarce public-sector funds, the shadow price of public money might itself rise more rapidly, and in these circumstances it is the private sector which enables construction to be brought forward.

The major problem with a purely market-based approach arises over the treatment of different modes of transport. Here there is a dual problem between, on the one hand, private and public transport and, on the other, those modes which use a publicly provided infrastructure as opposed to those responsible for their own track. One of the reasons for subsidising public transport, and especially public transport responsible for its own track (such as railways), has been that private transport using public infrastructure has not been asked to meet its full marginal social costs, especially in congested conditions. During periods when there are changes in the pattern of transport demand this problem is intensified. Rail will be left with the full costs of maintaining an old, underused infrastructure in a way that road users, private or public, are not. This increases its difficulty of being able to invest in new infrastructure in areas of rising demand.

Recent government policy in Britain has been based on the concept of the level playing-field (Rickard, 1990) which correctly argues that subsidies which distort the resource costs of competing modes are wrong. Hence all modes should compete on an equal basis, and government should not interfere to try to affect either individuals' choice of mode or global modal split. Unfortunately, this argument does not make provision in the short term for existing imbalances in resource costs. The argument for rail is not an argument for rail for its own sake, but an argument about efficiency in resource costs. One solution to this, which has already been implemented in some other countries — for example, Sweden, Switzerland and Austria — and which has been proposed at a European Community level, is the separation of track and service provision such that railway operation would compete on more equal terms with road operation.

The other major problem with the approach adopted so far arises in an international context. To some extent, consistency of transport planning parameters only becomes important when modes of transport have to meet across international frontiers. In this sense the confrontation of recent British transport policy with differing Continental practices has been heightened by the direct physical contact of rail which the Channel Tunnel will bring. Britain's interests have been served by introducing more competition into the operation of the inherently international modes of ferries and airlines, although there is a question over the degree of implicit subsidy given both to access modes and to traffic-control systems. Nevertheless, the British contribution to EC transport policy has been to introduce these sectors into the policy area and to press for deregulation. The problem with rail is that it is only peripherally international, in terms of the amount of traffic it carries, but that traffic can potentially affect the whole network. Furthermore, rail infrastructure impinges over a wider area than other public international modes.

However, there is another more fundamental point. The use of transport as an input means that excessive resource costs devoted to it imply a loss of competitiveness. A principal objective of completion of a single internal market in transport in Europe is to eliminate the hidden subsidies to industry which transport regulation and subsidy imply. This does not mean that Europe's competitive interests will be best served by ignoring the imbalance in marginal resource costs between different forms of transport. It is to this European dimension that we turn in the following section.

9.3 The market in the context of European transport planning

There are two basic themes which need to be addressed in this section. First is the question of alternative approaches to transport provision adopted in other European countries, and how these affect British interests. Second is the question of an emergent Europe-wide concern with transport, which is particularly relevant to new infrastructure of European significance.

There are many myths about transport provision and policy in Europe. From a British perspective other European railways are either seen as modern, technologically advanced and efficient, or as grossly oversubsidised. In fact, the notion that countries such as France and Germany are investing in rail to the exclusion of road is not borne out by the evidence, since both cater for higher levels of car ownership with larger networks of motorway-standard roads (see Bendixson (1989) for a detailed discussion). West Germany invested an average 0.79 per cent of its GDP each year in road infrastructure in the early 1980s and France 0.65 per cent, compared with the UK figure of 0.40 per cent. In the same period, investment in rail in Germany was 0.26 per cent of GDP, while in both France and the UK it was 0.09 per cent. Despite the commitment of a £20 billion investment plan for high-speed rail over 20 years, France is also committed to a 250 per cent increase in the size of its motorway network between 1987 and 1997.

What is, however, much more evident is the extent to which investment plans in continental Europe are part of an overall transport strategy which has links to regional development strategies. Furthermore, there is a greater emphasis in mainland Europe on planning the interface between modes of transport. Increasingly this is coming to mean links between rail as the dominant mode for inter-city business travel and, on the one hand, air travel and, on the other, intra-urban links. Thus a growing number of European airports are not just rail-linked, but are on main inter-city rail routes. New high-speed rail interchanges are being planned or built at Charles de Gaulle airport, Paris, and Satolas airport, Lyon, as part of the French TGV network, and similar plans also exist for Zaventem, Brussels, and Cologne–Bonn airport on the proposed Cologne–Frankfurt high-speed rail route.

The relevant question, therefore, is not concerned with total subsidy levels of different railway undertakings, but with how new investment is appraised and financed. Total subsidy levels are determined more by accumulated problems of network size, maintaining non-viable rural railways and coping with overmanning, than they are by new investment. Investment in new rail infrastructure is largely justified on the basis of benefits it will bring to the rest of the railway network, which means that it must generate an adequate rate of return.

SNCF looks for a positive net present value, over a 20-year period, equivalent to a rate of return of 12–13 per cent on appraisals for new TGV lines, within a financial structure laid down in a four-year *contrat de plant*[1] For investments meeting this criterion, SNCF undertakes the investment itself from its own resources, and by borrowing against government guarantees, as with the TGV Sud-Est and TGV Nord. Where initial appraisal does not meet this criterion, as with TGV Atlantique and improvements to the traditional rail network in Nord-Pas de Calais, SNCF looks for specific government grants of an amount sufficient to enable it to meet its target return. On TGV Atlantique, this amounted to a 30 per cent grant, justified on environmental protection and regional development grounds. For improvements in Nord-Pas de Calais, costs are being shared equally by SNCF, the French state and Regional Council under general provisions laid down in *contrats de plan*.

Where specific interest groups seek variations to SNCF's preferred route, as with the route of TGV Nord via a new station in the centre of Lille, the extra cost has to be met (in part at least) by the local authority. This does not mean that SNCF can be dictated to by any group — witness the failure of Amiens to secure a diversion of TGV Nord to serve the regional capital of Picardie — since an impartial judge ultimately rules on the balance of regional versus national interests before authorising the *déclaration d'utilité publique* which confers on SNCF the powers it needs to achieve construction and completion.

Furthermore, a rate of return of 12 per cent to SNCF does not mean that it is looking for a 50% greater return than BR needs to find with its minimum 8 per cent imposed by the Treasury, since the accounting practices of the railways differ. In particular, SNCF receives two large blanket subsidies, one towards equalising track costs of road and rail (10.3 billion francs in 1990) and one towards compensating it for state-imposed conditions of employment, especially pensions (13.9 billion francs in 1990). Both of these could be regarded as attempts to generate a level playing field. Most uneconomic services are provided through direct contracts. Increasingly these are between SNCF and the various regional councils, a solution similar to that which BR now has with the Passenger Transport Authorities in the former metropolitan counties, and which was recommended for more widespread use by the Monopolies and Mergers Commission (1989) in its report on BR's provincial services. The national subsidy to SNCF for regional services is not very different in cash terms from that received by BR, but it covers a much larger network.

Turning to the question of appraisal of new infrastructures, the first major issue is the extent to which comparisons are influenced by geography. French experience suggests that rail can start winning business travellers from air with city centre to city centre journey times of three hours, and that at two hours it can dominate the market. Paying for a new railway is not just about market share, it is about total market size. This implies that the cities joined must be of sufficient size and economic importance to generate a substantial volume of traffic, and also that access to a wider network of existing rail lines is needed. It is not clear that there is a magic minimum size of traffic flow, but flows of less than about 12–15 million passengers a year over a minimum distance of 300 km are unlikely to be prime candidates. This recognises that traffic has also to be abstracted from existing railway routes on which fares may remain lower. SNCF has imposed substantial premium fares on its TGV services, plus heavy peak surcharges.

TGV Sud-Est reveals itself as a perfect illustration of this argument. As was noted in Chapter 7, rail was carrying some 12 million passengers a year in 1980, prior to TGV, and air some 2.1 million, on the Paris–Lyon route. In just four years of operation, the TGV had secured nearly 15 million passengers. One-third of these were new rail passengers, of whom one-third switched from air (nearly 60 per cent of the air market). Only a little over one-third of all passengers were in the core Paris–Lyon market, the remainder (and the markets showing the strongest growth) were from the variety of routes sharing the main trunk line, but serving destinations such as Lausanne, Geneva and Marseille. The 417 km of new line enabled TGV trains to serve a 2500 km network. An operating surplus in excess of 60 per cent of revenue raised the internal rate of return to 15 per cent for SNCF, and it is estimated that the social return is about double this figure.

TGV Atlantique actually serves a larger market in total, 15 million passengers in 1980, but spread over a wider geographical area and with fewer major generators of traffic in less buoyant regional markets. To be anywhere near viable, the new route had to combine traffic flows to the west (Bretagne) with those to the south-west (Bordeaux and beyond). The new route involves only 283 km of new line, but the maximum distance travelled by any one service is only 228 km, compared with a maximum 390 km between Paris and Lyon.

TGV Nord is in some respects a more difficult proposition for SNCF, since it depends much more on international links. The total new route planned is around 330 km, but the key internal market, Lille, is only around 250 km from Paris, a journey of about two hours minimum at present (within the critical barrier for daily journeys), and much less buoyant. There is no real air market to tap into, except for passengers wishing to interline in a Paris airport. SNCF estimated a little under 9 million passengers a year between Paris and northern French destinations, but the international traffic would raise this to 24 million (see Table 7.2 for more detail).

Even more significant here is the complementary development of the Interconnexion route around Paris. This 106 km line will enable cross-Paris journeys between the three main TGV routes without the time-consuming connection between Paris terminal stations. It will also give access to Roissy-Charles de Gaulle airport from all three new lines. This has a more marginal rate of return of 10 per cent (its construction costs being 30 per cent higher than TGV Nord) but clearly has a critical role in enabling SNCF to exploit the rest of its new network to the full. Hence it is being financed internally.

This signals not increased competition between air and rail, but complementary developments of the most appropriate and efficient means of transport for each type of travel. Charles de Gaulle itself is planned to be expanded to five terminals and runways capable of handling 100 million passengers a year, five times its present level of traffic. Competition is thus with other European airports, particularly Heathrow, the current leading European air hub with a capacity of 38 million passengers a year (which is almost saturated). This demonstrates very clearly the competitive pressures developing at an international level, and the critical importance of a strategy for all modes of transport.

It is likely to run further than just competition between major airports, however, since these are just one factor in determining the competitiveness of Europe's cities. Access to and within major cities is also vital to their continuing economic importance, a factor which will be investigated further in Chapter 10. It is interesting to note how quickly Germany has responded to this in terms of identifying infrastructure requirements for the newly united state, and the need to integrate Berlin into a European network. Even before unification, plans were announced for a Hanover–Berlin high-speed rail route, and studies are well advanced for a Berlin–Munich link. Rail traffic between the two former German states is forecast to grow from 6 million journeys a year in 1985 to 43 million by 2010, 28 million on the northern link and 15 million on the southern link (Blum, 1991). This latter is of the same order as traffic forecast through the Channel Tunnel in the mid-1990s. It is even suggested that there is sufficient potential demand on the northern link to support three high-speed lines from Berlin, to Hamburg and Frankfurt as well as to Hanover (and Cologne).

The problem is how far this national interest in promoting transport for competitive reasons is consistent with development of a European network. Here there are strong pressures towards deregulation in airlines and road freight transport, which dominate international business passenger and freight movements, respectively, and which might appear to sit uneasily alongside the promotion of rail. Nevertheless, the Community of European Railways (CER), which brings together the railway companies of the 12 EC countries, plus those of Austria and Switzerland, has had direct encouragement from the EC. Thus, in a 1986 report, the Commission expressed its determination to develop 'a European high-speed [rail] network' (Commission of the European Communities, 1986b) (Report COM (86) 341 Final), and on 16 September 1987 the European Parliament adopted a resolution to this effect. As the CER notes, 'a [European high-speed rail] network would be a major factor in strengthening the links that bind the Community together' (Community of European Railways, 1989, p.3).

Table 9.1 *Development of a European high-speed rail network*

	High-speed lines (km) New	Upgraded	Total	Costs (bn ECU, 1985)
V1 (1995)	5200	7100	12300	43
V2 (2005)	7100	8400	15500	58
V3 (Long-term)	9100	9900	19000	90
Link lines			6500	
Feeder lines			4500	
Grand total			30000	

Source: Community of European Railways (1989)

In its 1989 Report, the CER estimated that high-speed rail projects then completed or in hand allowed for a total expenditure of some 15 billion ECU. Over a longer time-span it envisaged the development of the European high-speed rail network shown in Table 9.1 and Figure 9.1, with a total cost of 90 billion ECU.

Campaigns for transport investment in other sectors, such as air and road, are also beginning to be conceived in these terms. Thus, in June 1989 European airlines were reported to be asking the EC to provide an integrated European air traffic control system. A confidential survey prepared by the Association of European Airlines, which represents all the major carriers, was reported as arguing that 'There should be only one European air traffic control system bringing with it the concrete leadership that is needed to correct the flaws of the current system and implement a system for the future' (*The Independent*, 1989a). Not only pressure groups representing individual modes such as rail and air have been involved in this. Groups such as the Round Table of European Industrialists (1988) have also campaigned both for massive investment in all means of transport (32–40 billion ECU per year was seen as needed) and, perhaps more surprisingly, for a European Infrastructure Authority to co-ordinate plans and the required finance.

Parallel to its interests in a high-speed rail network, the European Commission has been interested in proposals for a medium-term transport infrastructure programme together with a means of providing some seedcorn finance through a new financial instrument (Commission of the European Communities, 1986a; 1988b). This was seen as vital in developing the single transport market within the 1992 programme. An initial programme costing 5–6 billion ECU, and implying Community funding of some 630 million ECU over five years, met with serious opposition and has been progressively scaled down. A revised, more modest, proposal was made in 1989 (Commmission of the European Communities, 1989) and the Council finally gave approval in November 1990 to a programme involving expenditure of 240 million ECU by 1992 (for further discussion see Vickerman, 1991a).

It is also relevant to note the synergy between the Channel Tunnel and

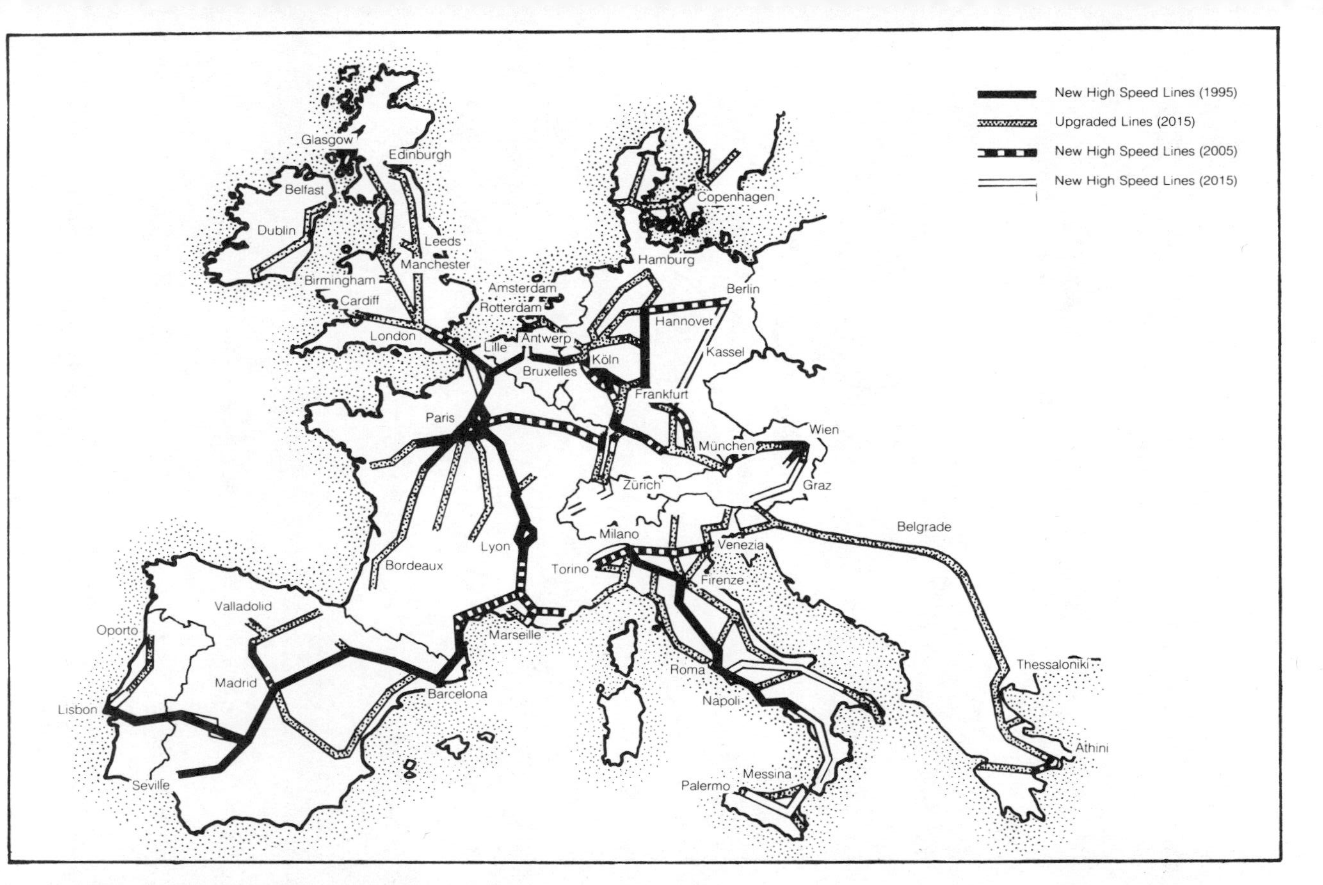

Figure 9.1 *European high-speed rail network*

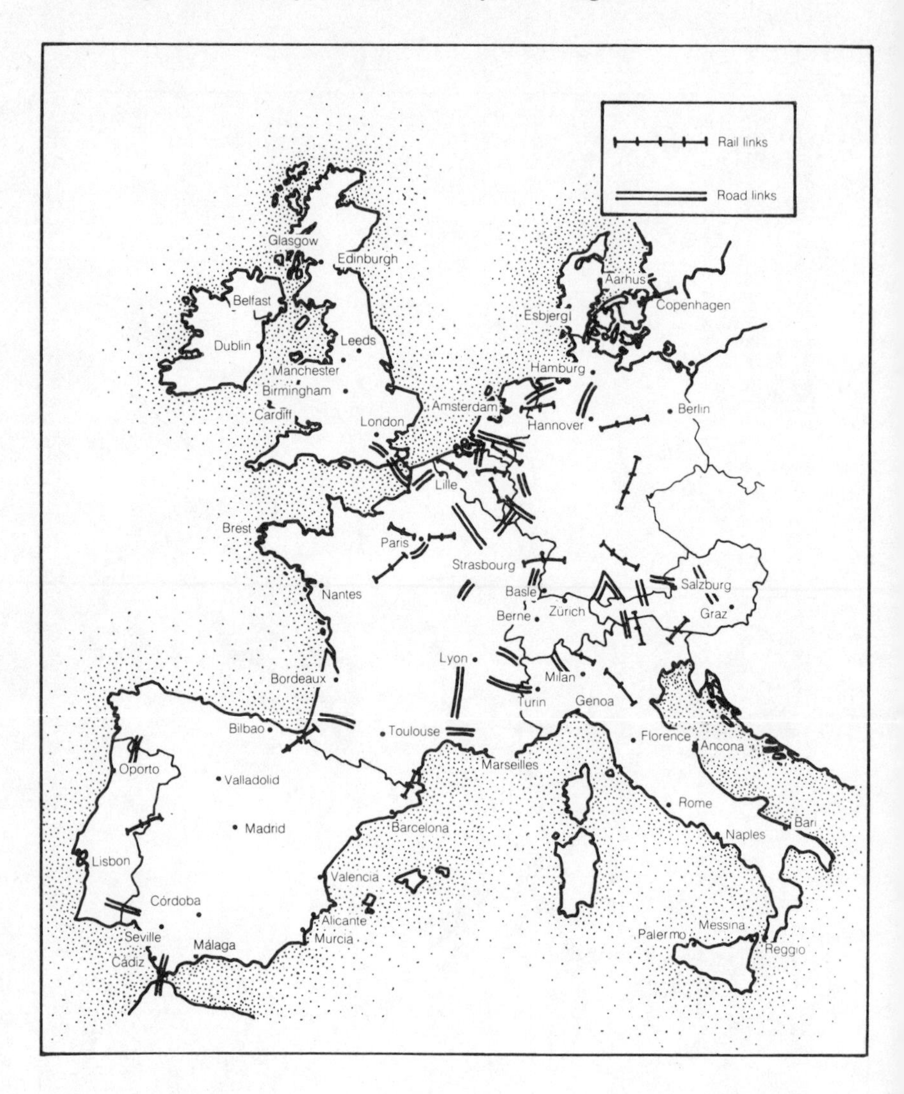

Figure 9.2 *Missing links in the European network*

other major European Plans to eliminate missing links, shown in Figure 9.2. Prominent among these are:

(1) The Great Belt crossing in Denmark, where a fixed link will be provided between the islands of Zealand and Funen, and thence to the Jutland peninsula. The 18 km link consists in part of a road-rail bridge, and in part of a road bridge and rail tunnel. The estimated total cost of this project, which is public-sector funded, was 2.25

billion ECUs (*c.* £1.5 billion) in 1989. The rail link should open in 1993, and the road link in 1996 (Illeris and Jakobsen, 1991).

(2) The Øresund crossing across the Sound between Malmö (Sweden) and Copenhagen (Denmark), which is being promoted by a consortium comprising 55 of the largest firms in Denmark, Finland, Norway and Sweden. Its initial ambitious plan was to build a four-lane motorway linking Oslo, Gothenburg, Malmö and Copenhagen; a double-track railway linking Oslo, Malmö and Copenhagen; and to invest in related infrastructure improvements (Walberg, 1989). Following the Swedish government's decision in favour in April 1990, it now seems that an 18 km road bridge, financed publicly, may be completed by 1998 at a cost of some £1.3 billion (*The Economist*, 1990).

(3) The Lolland crossing across the Fehmarn Sound which separates the southern Danish island of Lolland from north Germany. A bridge or tunnel crossing of this 24 km stretch is estimated to cost some £2–2.5 billion (*The Economist*, 1990).

(4) The Strait of Messina crossing between the toe of Calabria and Sicily, where a decision in principle in favour of a fixed link was taken by the Italian government in December 1985, and where the specific choice of a bridge was made in 1986. This £5 billion project is now awaiting firm proposals to be brought forward in 1992 (Secchi *et al.*, 1991).

(5) The Strait of Gibraltar crossing between Spain and Morocco, where two tunnel and two bridge projects have been proposed at a largely speculative cost of at least £5 billion.

(6) A new Alpine Rail Axis through Switzerland. Following a major study of alternatives, a compromise involving the development of both a western link via the Simplon–Lötschberg route, and an eastern link via the St Gotthard route has been proposed. The cost of some 7,600 million Swiss francs implies rather low rates of return (5–6 per cent maximum), but the links are seen as crucial to Swiss environmental policy in promoting the use of rail over road, for through traffic of relatively low benefit to the Swiss economy.

The key point here is the emphasis on rail in many of these schemes, which will enhance rail's competitiveness both through elimination of modal changes or bottlenecks, and through facilitation of long-distance rail haulage where rail has an economic advantage. This has particular significance for the more peripheral EC countries such as Italy, Spain and the UK. Each of these links depends, therefore, on every other link to enhance its financial and overall economic viability. From a European perspective, incremental appraisal is ideally required to prioritise between such investments. What is certainly clear is that they are too important to be left either to individual governments or the private sector.

It is significant that the only body able to take any real part in this development at a European level has been the European Investment Bank (EIB). The EIB significantly sought a major role in the finance of the

Channel Tunnel, despite the fact that it had to adopt unusual rules for both the initial loan of £1 billion (secured on commercial bank letters of credit rather than government guarantees) and the further loan of £300 million (secured on the existing assets of Eurotunnel). This makes EIB loans the largest single financial stake in the Channel Tunnel by a substantial margin, and reinforces the argument about the Tunnel's wider European significance.

In December 1990, EC Commissioner Karel van Miert presented a master plan for a 25 000 km network of new and upgraded railways to the EC Council of Ministers. This was costed at 150 billion ECU, of which only 60 million ECU would be found from the EC budget. The rest would be shared between member states and the private sector. Particularly relevant in view of the above discussion is the focus on 15 'key links', mainly located in border regions which tie together national plans for high-speed lines. One of these is the link between London and the Channel Tunnel.

9.4 Conclusions

This chapter has sought to show how the debate about privatisation needs to be widened to allow for consideration of both access links to the Channel Tunnel and how the Tunnel and its associated networks fit into an emerging European plan for transport infrastructure.

The effectiveness of privatisation of the Tunnel on efficiency is not yet clear. The construction phase has not been free of similar problems to those traditionally faced by large public-sector infrastructures. Regulators, such as the MMC, are clearly looking carefully at potential problems which might arise in the operation phase.

The transport policy debate in Britain has, however, been dominated by clear misunderstandings of what a 'level playing field' for transport should look like — this has affected the ability of rail to compete particularly seriously. Rail seems to be both the area where British policy is likely to come most directly into conflict with other European, particularly French, policy approaches, and the area where most firm plans for future European action are being formed. There is clearly room for considerable future conflict between London and Brussels over development and funding of a direct rail link between the two.

The Channel Tunnel itself is not, therefore, a positive planning disaster, but it may well have opened the door to a subsequent one of considerable magnitude.

Note

1. The minimum acceptable rate of return on SNCF investments is 8.2 per cent.

10 *The Channel Tunnel and regional economic development*

New transport infrastructure has important implications for economic development. However, it is extremely difficult to predict accurately in advance what those implications will be. Historically, the general presumption has been that although there is a clear association between the level of transport provision and a region's economic performance, this does not amount to a causal relationship (Straszheim, 1972; Gwilliam, 1979). Moreover studies which have tried to quantify the economic spin-off from particular infrastructures have found it difficult to separate out unambiguously the direct impact of the infrastructure from other concurrent changes (Botham, 1983). Hence, in the UK, for example, the regional economic consequences of road investment are systematically excluded from approved highway appraisal exercises (Department of Transport, 1977). Similarly, British Rail is only allowed to include benefits which can be turned directly into revenue in its appraisal of new investment (see Monopolies and Mergers Commission, 1988; 1989a).

Nevertheless, that there is an association between infrastructure and economic development is well documented in Europe (Biehl, 1986). Biehl (1991) argues that infrastructure is a public good which serves to raise the productivity of private resources in a region. Where there is excess capacity in infrastructure provision, however, a lack of other resources will impose a constraint on growth rates. The real importance of infrastructure therefore occurs when it creates a bottleneck. Blum (1982) has presented clear evidence of the way in which the absence of transport infrastructure at a local level can serve to reduce the actual income of an area below its theoretical potential. Hence infrastructure provision does not produce incremental effects, which accounts for the ambiguities discussed above.

In the case of the Channel Tunnel, private financing of the project has precluded detailed cost–benefit analysis of its wider impact. Some analysis has been undertaken at local and regional levels, but little analysis has been carried out of the wider regional, national or European implications, especially of indirect and induced effects.

This deficiency is particularly important in the UK, and at the wider

Table 10.1 *National distribution of net benefits from a Channel Tunnel, 1980 (per cent)*

France	47.0
UK	29.5
Belgium and Luxembourg	3.3
Netherlands	2.9
Federal Republic of Germany	2.8
Spain	1.9
Other countries	12.6

Source: European Parliament (1981)

European level; it is not so important in France. This is because in France the main impacts of the Tunnel, both positive and negative, are expected to be concentrated in Nord-Pas de Calais. Indeed, this is an explicit policy aim. In the UK, the regional economic impact of the Tunnel is expected to be more dispersed; similarly, in the EC as a whole — with the exception of Nord-Pas de Calais — Tunnel impacts are likely to be felt in a variety of regions. In these circumstances, the issues raised by the Tunnel in a particular region, set of regions, or country, are difficult to identify.

The purpose of this chapter is to go some way towards making such an identification. It is well beyond the scope of this book to present a comprehensive cost–benefit analysis of the regional economic impact of the Channel Tunnel. What can, however, be undertaken here is an analysis of the conditions of likely economic gain and loss from the project.

10.1 Existing studies

Precursors to the present Channel Tunnel project were, at various stages, subjected to limited cost–benefit analyses of transport effects (that is, impacts on non-users as well as on users of a fixed link) in 1963 (Ministry of Transport, 1963) and 1973 (Department of the Environment, 1973b; see also Gwilliam, 1983). These did not take into account wider regional effects. Similarly, the De Keersmaeker Report to the European Parliament (1981), based on the work of Coopers & Lybrand and SETEC Economie (Commission of the European Communities, 1980), only estimated the distribution effects of user benefits plus the multiplier effects of construction. The national distribution of net benefits (Table 10.1) cannot be seen as a full estimate of the spatial distribution of impacts.

The two regions adjacent to the tunnel did undertake wider analyses of the impact of the Tunnel, both on existing transport services (such as the ferry services which are major employers in their ports) and on the wider economies through indirect and induced effects (Conseil Régional

Nord-Pas de Calais, 1985; Channel Tunnel Joint Consultative Committee, 1987b). However, despite claims made about the beneficial effects of the Tunnel on the whole of the UK economy in the Channel Tunnel White Paper (Department of Transport, 1986a) and the Anglo-French treaty (HMSO, 1986), these were never substantiated.

The House of Commons Transport Committee, in looking at the question of the fixed link, argued that it was important to know what the impact on employment would be, but maintained that it was 'difficult to distinguish in which regions this will occur' (House of Commons, 1985, para. 118). The government's view that there should be no regional impact analysis was reaffirmed by the Select Committee on the Channel Tunnel Bill (House of Commons, 1986, para. 29). In response to specific requests from the North of England Regional Consortium, the then Minister (David Mitchell) replied that:

> the Government's view is that the impact of the Tunnel on the North of England will be positive rather than the adverse . . . we do not expect that the Tunnel will have a significant influence on the regional distribution of industry and employment . . . I do not believe that an analysis of the wider regional impact of the Tunnel would be a useful exercise or could sensibly be undertaken at this stage (House of Commons, 1986, Appendix 16).

Since there were in existence systems for monitoring the regional situation, he believed that these would provide sufficient information. This lack of government initiative has been critically reviewed in more detail by Vickerman (1988).

In the absence of an authoritative study of Tunnel impacts outside Kent and Nord-Pas de Calais, a number of studies have nevertheless been commissioned by UK regions which feel that the Tunnel is likely to have an important (negative) impact on their economies, and which wish to have an 'objective' or 'expert' study with which to lobby government. Included in this category are reports on the North of England (Centre for Local Economic Strategies, 1989), on the Scottish economy (Pieda, 1989), on London's business community (London Chamber of Commerce, 1989), on the South East Region (SERPLAN, 1989), on Wales (Atkins, 1989), on Cornwall (Cornwall County Council *et al.*, 1989) and by the Royal Town Planning Institute (1990). Most major regions of the UK and many local authority districts will, before long, have studied the likely impact of the Tunnel on their economies.

Rather less frenetic activity has occurred on the French side. The national regional planning agency, DATAR, commissioned a report on the four northern regions, Nord-Pas de Calais, Champagne-Ardenne, Picardie and Haute Normandie (Metge and Potel, 1987). However, the situation in France is substantially different in terms of the focus of regional effects, as has been argued in more detail elsewhere (Holliday and Vickerman, 1990).

Apart from the UK and France, the main interest has come from the Irish Republic, which fears its increasing isolation as an island and wishes to press for its own improved connectivity to mainland Europe via the Tunnel (Ferris, 1990). The European Parliament (1988b) took up the theme of the Tunnel's wider impact, and this has led to a major study on impacts throughout the Community being undertaken in 1990–1.

Each regional report in the UK has tended to address the almost simultaneous creation of a single European market at the end of 1992, and the opening of the Channel Tunnel in mid-1993. Typically, each event, and particularly the Channel Tunnel, is seen as a potential *threat* to regional economies. The key issue here is infrastructure links to the Tunnel, and in particular rail links. Without substantial investment in the national rail infrastructure, all British regions believe that they will be at a competitive disadvantage to the regions of mainland Europe, and that their peripherality within the EC will be increased. The key policy issue here is the problems posed by congestion in London and the South-East. However, most regional assessments tend to be based on assumptions of a static economic structure: the regional impact of changes in transport costs and competition are then estimated on this basis. With the limited exception of the *Kent Impact Study*, regional studies have tended to overlook the more dynamic changes to regional economies which 1992 and the Channel Tunnel can be expected to provoke. By ignoring potential industrial relocation and a likely restructuring of production, regional studies tend to be unduly pessimistic about the impact of 1992 and the Channel Tunnel on regional economies.

These regional studies have, however, changed the emphasis of the debate. At the start of the present project, investment in the Tunnel and associated infrastructure was suspected by distant UK regions of being merely a reinforcement of the North–South divide (and, therefore, a bad thing). Now other regions recognise that investment in infrastructure in the South-East is a major determinant of economic impacts beyond London (and is thus a good thing). It should be noted, again, that this point is also made at the EC level, where the Irish Republic argues that investment in road and rail links in the UK is vital to economic development in the Republic.

Thus there is now a substantial consensus in the UK regions concerning investment in UK rail infrastructure. The government has so far only been prepared to accept British Rail's modest programme of upgrading existing routes to carry tunnel passenger and freight traffic from 1993, financed out of BR's permitted investment programme. The key questions which it must address are, however, evident. They are when this capacity will become inadequate, and whether, even before saturation, improved quality of service would be beneficial to the regions. However, the speculative nature of any investment in a new high-speed line to answer such questions poses difficulties for BR's required 8 per cent rate of return on new investments, given its continuing need to incorporate the full environmental cost of a new line in the appraisal. The Thatcher government accepted the case for the eventual need, but argued that it should be wholly financed on a user-pays basis, preferably with a private-sector partner to take the investment outside BR's cash limits.

The regional lobby argues that rail links in Kent and around London should at least in part be publicly funded. The unity of interest between Kent and other regions of the UK was clearly demonstrated by a conference hosted by Kent County Council in Leeds in April 1990. The (then) Transport Minister, Michael Portillo, reiterated that the government would not subsidise rail links to the Tunnel. By contrast, the regions themselves agreed that public subsidy of rail infrastructure in Kent was necessary to the future well-being of all parts of the UK economy.

Thus, rail investment has come to stand at the centre of British debate about the regional impact of the Tunnel. The campaign for increased investment in road infrastructure appears, at present, to be a secondary debate, perhaps because of increased environmental awareness. Nevertheless, figures quoted by the Confederation of British Industry (1989b) — that transport congestion in the UK currently costs £15 billion a year — clearly highlight the problem.[1] A similar switch of attention is also evident in France (though it is not always directly linked to the Tunnel, in connection with which a major programme of road links is being completed) and, indeed, throughout the EC.

10.2 A basis for analysis

Central to any question of the regional consequences is the 'European' impact of the Channel Tunnel. Construction of the Tunnel will have an effect not only on the transport of goods and people between the UK and France, but also, because of its relationship to other infrastructure, on transport patterns over a wide area of north-west Europe. In the long run, transport changes will lead to potential changes in regional locational advantage, and hence to a possible redistribution of economic activity between regions. Outside Kent and Nord-Pas de Calais, these are, however, likely to be small in absolute terms and difficult to measure, but cumulatively of greater significance. In these circumstances, micro-economic analysis — of the impact of change in both passenger (business and tourist) and freight markets — becomes imperative if the European impact of the Channel Tunnel is to be understood. Clearly, this analysis cannot be undertaken within the scope of this book, but the basis on which it might proceed can be given.

Infrastructure can have substantially different effects according to its context. Vickerman (1991b) argues that there are three basic ways in which transport infrastructure might affect the economic development of a region. In the first, infrastructure simply frees the passage of traffic through the region, without its having much impact. This would be the case of motorways with few access points, or high-speed railways with no stations. This may be termed a *pure corridor effect*. In the second, new infrastructure can improve communications into and out of a region. Such improvements could be in just one direction, typically for a peripheral region, or more importantly in several directions, in which case a *crossroads effect* will be felt. In the third, the infrastructure may improve communications within a region, generating changes in the internal efficiency of the region, in which case it will have an *internal efficiency effect*.

Any given new infrastructure may have any one, or any combination, of these effects on traffic flows. Understanding their interrelationship is critical to understanding the possibilities for development in a region. However, there is also a further dimension to be considered, which is the impact of infrastructure investment in one region on economic development in another. Such impact may be direct or indirect.

Indirect effects derive from changes in the relative competitiveness of regions as a result of infrastructure investment. Thus, the creation of a new link into a region enhances that region's potential relative to a similarly located region which is left without such a link. This effect is demonstrated by the clamour of French regions to be placed on the TGV network (CEDRE, 1990). This is an extension of the argument that lack of transport acts as a constraint on development.

Direct effects involve both the objectively measurable impacts on transport costs faced by particular industries and the subjective views of the relative position of particular regions. Measurable impacts depend on changes in transport costs for specific routes, including both the direct impact of the new infrastructure and access routes to that infrastructure. Hence, the Channel Tunnel could have major impacts on distant regions. These impacts may include both reductions in costs associated with the new infrastructure, and potential increases in costs associated with greater congestion on key access routes.

The perception of these changes is more difficult to measure, since existing evidence from infrastructure investment seems not to be entirely relevant to the rather more fundamental changes implied by a major project such as the Channel Tunnel. We do have some knowledge of both shipper and passenger response to price and quality changes. The difficulty is in determining their transferability.

Subjective assessments of regions as a whole may be altered by a major project. The cost of crossing the Channel will not change that much as a result of Channel Tunnel operation, but the Tunnel has clearly already had a positive impact on the relative isolation of the UK within the EC. This could easily have an impact on investment decisions. Indeed, it is clear from existing research that perceptions of regions' characteristics are better determinants of location decisions than are actual values of the same set of variables (Henley *et al.*, 1989).

The major impacts are on the logistic structure of product and transport chains, with consequent implications for the location of activities. These relocations of economic activity in Europe are often argued to be likely to lead towards greater concentration in core regions. According to this view, the first sector likely to be affected by such a process is distribution. The counter-view is that improved communications will benefit lower-cost peripheral regions by removing transport bottlenecks, and enable new activities to locate in them without a loss of logistic coherence. The balance between these two sets of pressures will depend on whether current investment decisions in particular regions mainly involve the expansion of existing firms, or new inward investment involving either branch plants or relocation. Also important are the relation of the new plant's activities to other plants in the same organisation, and whether the new investment is using different technology from existing plants in the same industry. Thus, the central question is what particular economic advantage is being exploited. To examine more closely the regions likely to gain and lose a typology of regions has been proposed (Vickerman, 1991b).

10.3 A typology of regions

A typology of regions for understanding the impact of the Channel Tunnel needs to reflect three key issues: geographical distance from the Tunnel, and hence the size of the immediate impact; functional position of the region within both its own national context and the wider EC context; and eligibility of a region for assistance from national regional policy, or from EC Structural Funds. This suggests that five basic regional types will be needed for most studies (although there may be a direct overlap between some of these in some cases). The five regional types are:

(1) Those containing the new infrastructure, which receive the direct impact of construction and of changes in accessibility. In the case of the Channel Tunnel, these are Kent and Nord-Pas de Calais.
(2) Other regions which lie between the main conurbation areas of north-west Europe (Greater London, Ile de France, Rhein–Main, Rhein–Ruhr, Randstad). Included here are other regions of northern France, most of Belgium (excluding Brussels and Antwerp), Limburg, Noord-Brabant and Zeeland (Netherlands). These regions stand out clearly as having below average GDP per capita in Figure 4.3.
(3) The main conurbation areas in the economic and geographical core of Europe. Included here are cities such as London, Paris, Brussels, Frankfurt, Cologne, Düsseldorf, Amsterdam and Rotterdam.
(4) The fast-growing new industrial regions of Europe which lie in the main just outside the core regions of Europe, including some of those which have benefited most from the growth of new technologies. Included here are regions such as the Western part of the Outer South-East of England, East Anglia, the East Midlands, Rhône-Alpes and Franche-Comté (France), Stuttgart (Germany), Lombardia (Italy).
(5) Peripheral regions, which will include the generally poorer regions of the EC, but which may nevertheless benefit from any general improvement in transport provision across the Community as a whole. Included in this category are the inner periphery such as Wales, Scotland, Bretagne, Auvergne and Emilia Romagna, and the outer periphery such as Northern Ireland, the Republic of Ireland,

and regions of central and southern Italy, Spain, Portugal and Greece.

Although distance from the Channel Tunnel is clearly a factor in generating this typology, it is important to note that distance itself is neither the sole nor, indeed, the major factor determining its impact. Rather, the functional position of a region in both national and European regional systems is critical. It is as important to understand economic interactions between regions as to understand the economy of each region in turn. This explains the generally non-contiguous nature of some of these regional groups, especially the functionally more independent areas of Types 3 and 4. Failure to develop this understanding is one of the main limitations of more conventional approaches. To overcome this limitation requires careful examination of the role of transport at a disaggregated, micro-economic level.

Type 1 regions will gain a specific direct impact from the new infrastructure. This starts with the direct impact of construction which has been significant in terms of employment in both Kent and Nord-Pas de Calais. These regions also face the threat of unemployment in traditional port and ferry industries, and the potential for environmental damage, if not from the Tunnel itself then at least from an increase in traffic and the development of connecting infrastructure. As we have seen in more detail in Chapter 6, the potential for creating new employment should not be exaggerated. These regions are both functionally dependent. Kent is close to, and dominated by, London and has relatively little dynamic industry. Nord-Pas de Calais is far enough from Paris not to be dominated by commuting but its industries are generally old and it displays all the characteristics of a functionally dependent declining industrial region, including substantial regional aid. The danger for both regions is that the Tunnel and its associated infrastructure simply ease the path of traffic between our Type 3 regions without their being able to establish a new dynamism.

Type 2 regions are the remaining regions which are particularly interesting, because they lie within the geographic core of Europe without being fully part of its economic core. They are essentially functionally dependent regions within their own countries. The nature of this dependency differs in kind and degree between regions and national contexts, being determined by industrial structure, location relative to neighbouring conurbations, and so on. These regions are relatively close to one another geographically, but have little interdependence between their economies, because of the proximity of national frontiers. They are more likely to interact with Type 3 regions on which they are dependent. Several parts of these regions are partially eligible for EC assistance under Objective 2 (declining industrial regions). Unlike Kent and Nord-Pas de Calais, they do not receive a direct boost from Tunnel construction and operation. Some, like West Vlaanderen, may receive substantially negative impacts due to a switch of traffic from traditional sea routes; others, like Hainault, may benefit from the rerouteing of some traffic, but lose out from the increased isolation promoted by the corridor effect of the new TGV line; while others, like Zeeland, may be left relatively untouched.

The conurbation regions of Type 3 are core regions of the European economy. They are its main generators of growth, and have shown increasing interdependence in recent years. This interdependence makes them reliant on fast and efficient transport, especially passenger, given the dominance of service sectors in their economies. Most other regions are functionally dependent on these regions, usually within a national context. The question which then arises is whether the removal of internal frontiers after 1992 will alter this pattern of dependency. Although the proportional significance on the internal economies of these regions of any new infrastructure built outside them may be small, the absolute benefit which they secure may well be great given their existing dominance. A critical policy question may then be how to divert some potential benefits away from these regions to other types of region. By improving accessibility across or around major cities, for example, their typical dominance over internal communications may be reduced. As new transport infrastructure reduces the importance of both national frontiers and physical barriers (such as the Channel) to communication, conurbations themselves may become more important traffic bottlenecks. The position of London is particularly relevant in this context. Conurbation regions are also dependent on the efficiency of their own internal transport systems and how they connect to national and international networks. The possibility of interchange between modes, for both passengers and freight, is critical here. Increasingly we are seeing the importance of interface between air and rail networks by connecting major airports to both local and inter-city rail networks. Cities such as Paris, Brussels, Amsterdam and Frankfurt will all eventually be served by long-distance trains using new high-speed lines, as well as by regional services.

Type 4 regions are in many respects Europe's most successful regions in terms of growth, and in particular in terms of growth of new, especially high-technology, industries. This growth has tended to release these regions from their former functional dependence on the major conurbations, and has placed them on a more equal footing with them. Furthermore, their interaction with each other has increased as a result of their shared technological bases (see, for example, Laboratoire d'Economie des Transports/Interalp, 1986). Locationally, these regions are often similar to Type 2 regions relative to the major conurbations, except that they suffer less from the border region effects of internal frontiers and from the spill-over effects of congestion. They stand to benefit from improved communications integrating them further into the 'core' of European regions. Indeed some of the main improvements to existing networks have affected these regions, examples being the M25 in south-east England, the TGV Sud-Est to Lyon, and the Mannheim–Stuttgart Neubaustrecke of the Deutsche Bundesbahn. The very success of these infrastructures has been remarkable, as is demonstrated by congestion on the M25, and growth of traffic on TGV Sud-Est. These regions are in the main not eligible for EC assistance, although they include some areas which are partially eligible under both Objectives 2 (declining industrial regions) and 5B (rural problem regions).

The peripheral regions of Type 5 are likely to show dependency at both national and European levels. They are a much less homogenous grouping than the other regional types, and include industrial peripheral regions in countries such as the UK, and poor agricultural regions in Italy, Spain and Portugal. These regions already suffer from peripherality, although its degree, and the intensity of the regional problem, differ within the group between, for example, the North of England, Ireland and the more remote parts of Portugal or Greece. The question which arises here is whether improved communications will reduce or intensify peripherality. UK regions, for example, already fear intensification as a result of the Channel Tunnel (Centre for Local Economic Strategies, 1989), but a more dynamic (and less pessimistic) alternative is possible (Vickerman and Flowerdew, 1990). These regions generally have lower unit costs than the other types of region, and this could lead to their use as low-cost production centres for activities which are decentralised from core regions of the Community. Their peripherality has not prevented rapid investment in recent years, often decentralising from core regions. The extent to which this continues to happen will, however, depend on the improvement in transport links. Lack of Tunnel-related infrastructure investment will act as a constraint on growth even if the provision of new transport links is not in itself a promoter of regional growth. There is already evidence of this tendency for Wales and some Spanish regions are suffering from uncertainty over future investment in converting the railways to standard gauge. Obvious sectors involved here are process and assembly industries, such as automobiles and foodstuffs.

An interesting corollary is that within the single European market after 1992, Type 2 regions are likely to be in competition with Type 5 regions for industrial development. Type 5 regions are all eligible for EC assistance, variously under Objectives 1 (structurally deficient regions), 2 (declining industrial regions) and 5B (rural problem regions). This illustrates the heterogeneity of such peripheral regions, which comprise both industrial regions in the UK, structurally backward regions in Ireland and southern Europe, and predominantly rural regions in France, the UK and parts of Germany and Denmark.

The message of this section for regions contemplating the effects of the Channel Tunnel is two-fold. First, there are many diverse influences on regional fortunes which need to be taken into account. Transport is important as a potential future bottleneck which could act to constrain continuing growth. Regions are, therefore, right to be concerned about maintaining and improving connectivity to a network. Second, however, the degree of competition between regions, both within and between countries, can be noted. It is this which makes some national governments, notably that of the UK, suspicious of the use of transport as an agent of regional growth. When there is a period of substantial change in functional relationships between regions, however, transport can be of greater potential importance as a diverter of new activity. We shall clearly be in such a period of change during the 1990s.

10.4 Infrastructure and changes in urbanisation

There is an important further dimension to the regional consequences of new infrastructure. Transport changes also lead to changes in settlement patterns and in the structure of urbanisation. Improved rail services in the UK have, for example, led to increases in long-distance commuting, compensating the rapid rise in house prices in the London area. Similar effects can be identified in the Paris region as a result of the expansion of the RER regional rail network. The ability of a city to expand through infrastructure investment helps contain rising costs in its centre, which in turn dampens the equilibrating effect which such rising costs would otherwise have through the diversion of economic activity to other regions (Klaassen, 1991). To some extent, the lack of investment in London's transport system can be seen as a deliberate policy attempt to secure a regional policy effect in the UK. The danger is that a lack of new investment in Channel Tunnel-related infrastructure as a continuation of this policy may have serious negative consequences. Although rising costs may make London less attractive than other UK cities, such as Birmingham, Manchester and Newcastle, they also make it less attractive than other European cities, such as Paris, Brussels and Frankfurt. Moreover, real competition is between cities of the same type or functional significance, and not between cities within a functional hierarchy. Thus, if London becomes less competitive at a European level, then not only London but also all the cities and regions which are functionally dependent on it will become less competitive.

The link through infrastructure to competitiveness, and the way it is transmitted through a spatial hierarchy, is a key issue on which we so far have only patchy and partial evidence.

There is, however, a further issue concerning access to networks, and the critical role it plays in regional development. The role which access points to networks play in determining their regional and local impact is well known in the case of motorways (Bonnafous *et al.*, 1975). More recent evidence from the French TGV confirms earlier observations that it is not just access to a network which is important in determining regional impact but also the way in which supporting infrastructure and local policies complement it (Laboratoire d'Economie des Transports/Interalp, 1986). Increasingly, however, transport, indeed whole logistical systems, are becoming multi-modal. The traditional argument about the economic development of major ports suggested that they benefited from their role as trans-shipment points. This is no longer the case, with unitised freight dominating, particularly intra-EC trade. Little or no economic development has been associated with Britain's two fastest-growing ports, Dover and Felixstowe. What multi-modal transport systems do is create mini-ports at strategic locations: the term 'freight villages' has been coined in the UK. Access to these multi-modal termini is seen as critical for regional development, since absence of such access will reduce the reliability of any service.

Such transport systems are important not so much for the carrying of finished goods, but as part of the production process itself, carrying intermediate goods and raw materials. Here, reliability is of the utmost importance to ensure 'just in time' delivery of inputs to users. Often, only a fully integrated transport system can achieve this. However, the more peripheral a region is, the more it relies on the efficiency of transport infrastructure in other regions to generate this reliability. Peripheral regions are attractive to potential investors because they have lower costs of inputs other than transport which can overcome the greater transport cost. Indeed, the higher direct cost of transport which arises from greater distances may also be offset by lower congestion levels on access to the primary network, and by possibilities for using more efficient transport systems for long hauls. This is one way in which more peripheral regions may be competitive with the more central, but non-metropolitan, regions closer to the Channel Tunnel.

10.5 The impact of infrastructure on industry

Discussion in preceding parts of this chapter has shown the great complexity of assessing the wide impact of transport infrastructure investment on regions as a whole. In this section we shall look more specifically at the impact on a more micro level on industrial sectors and firms.

The first task is to separate out pure transport effects from wider economic effects. The opening of the Tunnel will provoke both mode and route changes, and new trip generation. These pure transport impacts relate to the existing distribution of activities, and will affect, in particular, regions with ports and airports in the UK and on the Continent through which trade passes. Such changes will not have a profound knock-on effect on other employment in these regions, since most trade is unitised and passes straight through them. They will also have minimal effects on more distant regions.

However, the creation of a new piece of infrastructure with different characteristics from existing links can affect production and location decisions over a much wider area, making the impact of the Tunnel much more significant. Here, changes which only affect market areas, and do not affect location, must be distinguished from changes which lead to relocation. Changes which affect market areas are a consequence of shifts in competitiveness which result from differential changes in transport costs (Vickerman, 1989b). These are likely to benefit dynamic firms in fast-growing regions dominated by high technology, and firms with substantial existing scale economies in existing major metropolitan regions. The proportional impact on these regions may be small, but, as was noted above, in absolute terms it could be substantial. The likely result would then be a diversion of activity from firms in peripheral regions. This centripetal effect is one of the most feared impacts of the Channel Tunnel.

However, this is only part of the Tunnel's regional economic impact, since relocation may also be generated by new business opportunities created by it. Such relocation is unlikely to be induced solely by transport changes, though such changes enable a desired relocation to take into account a wider range of potential destinations, and to attain a closer approximation to an optimal location. This reorientation may particularly affect small and medium-sized enterprises (SMEs), which currently trade much less than larger companies since they face proportionately greater costs from existing barriers to cross-border trade. New transport opportunities in multi-modal transport which help minimise these barriers may make such firms more open to change than existing traders. The SME sector is also believed to be the one in which many of the impacts of the EC's 1992 programme will be most significantly felt (see Cappellin, 1990).

It is also crucial to take account of regional changes already in progress independently of any changes in transport provision. This is particularly true of restructuring occasioned by moves towards the single European market scheduled for completion in 1992. These changes may prove to be much more significant *per se* than changes in transport infrastructure, though there is also likely to be an important synergy between the two. In this context, particular attention should be paid to those sectors where transport delays and other border costs currently appear significant, and where significant regional restructuring as a result of scale economies or other dynamic changes to industry is likely to occur.

Certain key sectors may be particularly affected. These include major growth sectors, and sectors where transport is either a significant part of total costs — for example, transport of bulk materials — or where reductions in transport time could lead to significant gains — for example, in transporting perishable foodstuffs.

Three types of sectoral response can be identified. First, it is important to consider the extent to which changes in transport provision induce a relocation of economic activity between regions. Activities which come into this category will be footloose and sensitive to changes in the transport sector. They will, therefore, either have high levels of transport cost relative to total costs, and be seeking to reduce these, or be sensitive to the need to relocate relatively close to markets. Some traditional manufacturing industries might fall in the first group, though few industries have transport costs greater than 10 per cent of total costs. The second group could include most of the service sector, including high-technology industries, and business and financial services, where it is the movement of people, rather than of goods or materials, which is relevant.

Secondly, locally based industries which serve wider markets will be affected by the changes, inducing, for example, a faster rate of growth in sectors already important in the local economy.

The third type of activity is that with a primarily local orientation. While each of the first two types will depend on trade with groups outside the local region, this third type of activity depends entirely on the local market, and is therefore dependent on so-called primary or economic-base activities. Retailing and local business services fall into this group, but it can also include small and medium-sized construction companies, transport, public utilities and some manufacturing. These groups depend on the strength of the base sector for their success, but they are also vulnerable to some extent to outside competition. Hence, local retailing stands to face competition from new out-of-town developments by large multiples, local construction firms face competition from national firms for large contracts, and the nature of urban development in an area may change.

However, it must be recognised that transport costs account for a very small fraction of the total costs (value-added) of most manufacturing firms, typically less than 4 per cent. Hence a very large change in transport costs is required even to have an impact of 1 per cent on total costs of production. Furthermore, the larger the haul in question, the less will be the impact of any reduction in costs on one part of the haul. For example, the cost of the cross-Channel element will typically be between 15 and 40 per cent of the cost of moving a lorry between the UK and most Continental destinations. Since Eurotunnel claims a possible average 10 per cent cost reduction in real terms for the Channel crossing, this would give an average 0.01 per cent reduction in total costs.

This suggests that infrastructure like the Channel Tunnel will be insignificant as an agent of relocation. This, however, may be an over-simplification. Three main factors must be cited in correction of such a view. First, transport may be a more variable cost between locations than any other cost, and hence a substantial contributor to variations in profits. In itself, transport change is unlikely to generate a decision to relocate, but firms seeking relocation or expansion may take transport factors into account when choosing between alternative sites. Second, the absolute level of transport cost may be less important than transport reliability. If transport costs are 4 per cent of total costs, they may be as much as 40 per cent or more of profits, such that even a 10 per cent variation in transport costs is effectively a 4 per cent variation in profits. Third, lack of reliability in transport may generate greater costs than just the direct financial impact, since lost or late deliveries can halt a production process or lead to lost future orders. The size of costly inventories — it has been estimated that one week's inventories in the retailing sector cost some £200 million a year (CBI, 1989a) — may also reflect transport unreliability.

It is important to consider how transport can be traded off against other input costs and the level of output. Traditional theories tend to assume a level of output based on the pattern of demand, and then attempt to allocate this in space according to the cost of other inputs.

Transport simply enters as a means of trading off different locations on the basis of given amounts of input. A more complete theory of location requires that we allow the technology of production to vary across space. Hence firms may seek peripheral locations with higher transport costs if they can find cheaper inputs in those locations, but this may lead to a different technology of production in peripheral locations from those in core locations (Vickerman, 1989b).

This process becomes increasingly important in an integrated European market. The Cecchini Report has shown clearly that some of the major sources of economic gain within a single European market arise from the search for scale economies (Pratten, 1988). The question which then arises is whether this search for scale economies leads to larger plants, to larger firms, or to both, and whether this implies an expansion of existing plants or a search for a new technology, and therefore new plants which could be found in new locations. It is this part of the Cecchini study which is perhaps most speculative and most open to controversy, since it has implications for both industrial concentration and, potentially, geographical concentration.

Simplistic critiques assume greater concentration on both these dimensions and produce generally negative results for older and peripheral regions (Neuberger, 1989; Cutler *et al.*, 1989), but this seems to ignore two important factors. First, changing markets and the search for scale economies may lead to different 'best practices' in production. Second, the removal of barriers to trade associated with internal frontiers of the EC, and the harmonisation of technical standards, both allow for a greater substitution of transport for other factors of production. Hence, emphasis should not be placed simply on shorter frontier delays and consequent marginally lower transport costs. Instead, a more dynamic perspective should be adopted, in which the greater reliability of transport results both in greater cost savings and in integrated production processes which are less concentrated geographically.

Transport needs also vary according to what is actually being carried. The trend which can be seen with integration towards the single market is less towards the carrying of finished manufactured products and more towards the carriage of intermediate goods and inputs to industry. Since finished goods may have much higher bulk or weight, they may actually have a much higher transport cost than the carriage of intermediate goods. Thus greater dispersion of different stages of the manufacturing process need not be reflected in increased total transport costs for a manufacturing process as a whole. Reinforcing this is the way in which congestion in core regions of the Community is leading to rising costs of other major inputs, such as labour and land.

Analysis of sectoral change presents a rather mixed picture of possible future trends in the local economies of core regions of the Community. Some major positive points can certainly be identified, but it is clear that these regions will have to compete both with existing dynamic regions to attract high-technology industry, and with peripheral regions to attract industry of all kinds (Vickerman, 1990).

10.6 Centralisation or decentralisation in Europe in the 1990s?

Perhaps one of the major unresolved issues in the process of economic integration in Europe is whether it will result in an increasing concentration of economic activity in core regions of the Community, or whether it will provide a better basis for a renewal of the periphery. As is evident from the preceding analysis, the traditional view of a golden triangle at the centre of the Community, usually said to embrace the major conurbations of London, Paris and Frankfurt/Rhein–Ruhr, is beginning to break down. On the one hand, there is a problem in holding the core of Europe to be a single geographical entity since, in an economic sense, the core depends on industrial and commercial sectors, which are not necessarily associated with specific locations. On the other, there has been a movement out of the traditional manufacturing industrial sectors towards newer industries, including business and financial services. As a result, a new concept of the core has come to overlay the traditional, regional concept, one which stresses the increasing division between major metropolitan and other areas.

The economic core of Europe can be defined in a number of ways. The distribution of population density defines fairly clearly a core area of urban concentrations stretching from London across to the Rhein–Ruhr area of Germany, up the Rhein valley to Frankfurt and across to Paris. This includes the major urban concentrations of the Low Countries. Outside this central urban concentration are a number of important free-standing urban areas, including the major urban areas of industrial Britain, the large conurbations of Hamburg and Berlin, and the fast-growing centres of southern Germany, northern Italy and the Rhône-Alpes region. A second definition of the core is in terms of the major concentrations of manufacturing output within the Community. This tends to identify a slightly different area, stretching from the industrial heartlands of the UK down through eastern France and the western part of Germany as far as northern Italy. This is the region that has recently been dubbed the 'Blue Banana' (Brunet, 1989). A third definition looks at the economic strengths of particular areas, measured in terms of GDP or output per capita. This reveals a more fragmented definition of the core, incorporating characteristics of both population distribution and industrial activity. The main urban centres in the central areas are included, as are some of the free-standing cities, but many of the secondary urban areas in the centre are not. These urban areas of northern Britain, of Belgium and of parts of northern and eastern France, which depend on a traditional heavy industrial base, are increasingly badly equipped to face the demands of the modern era.

Cities like Lyon, Stuttgart and Munich, based on newer industries, have all experienced faster rates of growth than the traditional core, and these are the cities which have also begun to demonstrate major economic power in terms of output per capita. A more useful definition of the core needs to take into account not just physical location, but also functional position in terms not just of geographical location but also of ways in which regional economies relate to one another. Such relations depend on distance, but also on the actual needs of regional economies, in terms of transport of different types of materials or goods.

This recognises the relationship between core and periphery identified by Myrdal (1957), that as peripheral regions start to develop and to adopt new technologies and new industries from core regions, they increase demand for products which can only be produced in core regions of an economy. Instead of equilibrating the economic situation, these demands thus reinforce the rate of growth of core regions. This so-called 'backwash effect' therefore ensures the continuing pre-eminence of the economic core.

This view contrasts with the work of some geographers, such as Keeble *et al.* (1981; 1988), which has emphasised simple aggregate measures of economic potential, looking at income levels and the deterrent effect of distance or of transport costs as a means of defining the difference in accessibility of different regions. This is an inadequate approach, because it fails to recognise both the economic relationships that are involved, and indeed the way in which economic relationships may change through time as a result of technological change and of changes in the pattern of demand.

The picture in Europe is particularly problematic because the traditional economic core regions of the Community also happen to be located near to the geographic centre of Western Europe. However, if we look more closely at the current pattern of economic activity, we find that the geographic core has fragmented, and is represented by major urban areas, and not by the whole region, much of which is rather depressed in terms of output per capita. This process has been reinforced by the changing technology of transport which, for these new industries, has placed greater emphasis on speed and reliability.

The growing importance of air communications, which skip over intervening distances and provide a magnet for economic activity only around airports, has reinforced the dominance of the metropolitan regions. It has also allowed the core to incorporate newer and faster growing areas outside itself, with new industries looking for a better environment, and for locations away from areas associated with industrial decline. The development of high-speed railways is likely to reinforce this fragmentation and to increase further the dominance of metropolitan regions.

There is, however, a further factor which needs to be recognised in understanding the process of change. Traditional industrial economic activity was essentially constrained by the presence of raw materials, such as iron and coal, which are costly to transport. Modern industries are much less constrained. Furthermore, traditional industrial activity in Europe depended either on national markets or on overseas markets that could be served cheaply and easily by sea transport. It depended much less on trade between similar industrial countries. The remarkable feature of the postwar period and the process of economic integration in Europe is the growth of intra-European trade as a result of the creation of the

European Economic Community, and also of the role of intra-industry trade (Balassa and Bauwens, 1988).

The pattern of change can be associated with three major processes: first, change in the importance of different industrial sectors; second, change in transport provision; and third, expansion of the Community itself. New industries have developed in new locations away from the constraints of traditional industrial areas. Such industries have sought to escape the effects of congestion, of environmental decay and of the impact of traditional working practices on labour costs. While for some sectors this has involved an attempt to create new cores in new and dynamic locations, for more traditional manufacturing industries it has also involved a process of moving outwards into more peripheral regions, where lower-cost production facilities, which can be linked to existing plants within core regions, may be found.

Such development has itself both been occasioned by, and has necessitated changes in, surface transport. The growth of unitised trade, of roll-on roll-off ferries, of larger lorries and of motorway networks within the Community, has contributed to the possibilities of decentralisation. The growth of high-speed rail transport is a further means of incorporating more peripheral areas into a unified manufacturing process system. This is not just of benefit to passengers. It also enables capacity on existing rail lines to be freed for freight transport, thereby improving both its speed and its reliability.

However, the creation and expansion of the EC must be regarded as one of the major influences on the changing nature of the European core. Work in Germany by Bröcker and Peschel (1988) has shown quite clearly the impacts of integration, and their geographical distribution, not necessarily in geographically central areas, but in regions with important trading links. The major process in the 1980s has been the move westward following the accession of Spain and Portugal to the Community, reinforced by changes in transport, growth of new industries, and policy intervention. In addition, there is the possibility of a new axis of development along the Mediterranean coast, a European sun-belt, intersecting the traditional Manchester–Milan axis of industrial activity (Brunet, 1989).

The major change confronting Europe in the 1990s is the opening up of Eastern Europe, which will have an impact on the economic and geographical cores of Europe. The extent of this impact will depend on the extent to which these new economies are connected into the core of Europe. The impact on the existing core is, however, likely to be similar to that which has already been outlined for the development of the Mediterranean sun-belt. Again, there are severe dangers for those regions that are not part of the urban core.

Changes in transport underline these dangers. New forms of surface transport — such as high-speed passenger and freight railways, and limited-access motorways — serve to reinforce the position of major urban centres. If regions in the geographic core of Europe are not to be simply transit corridors, substantial policy initiatives may have to be developed.

Following the redefinition of assisted regions in Europe into the various 'objectives' of the structural funds, the European Commission has now embarked on a more ambitious plan to define a series of macro-regions in the Community. It is likely to be in the context of these macro-regions that new policy initiatives will be formed. Already regions in these groups are identifying common interests and forming collective voices. The Conference of Regions of North West Europe (CRONWE), for example, includes, Nord-Pas de Calais, Belgian and Dutch regions, Nordrhein-Westfalen and, more recently, Kent. Similar organisations cover the western and eastern Alpine regions (including non-EC regions in Switzerland, Austria, Yugoslavia and Hungary), and the Atlantic Arc. Transport questions have been much to the fore in many of their discussions (CRONWE, 1989; Cagnato, 1990).

10.7 Policy implications

The question of regional development is thus undergoing substantial change in the EC at the moment. Transport questions are assuming a new importance as the Community is moving to greater integration and the relevance of traditional policy measures is called into question. The first major consideration, therefore, in policy formulation is to understand clearly the level at which it is to be conceived and implemented: is it to be local, regional, national or Community? Three main issues which underlie this consideration have been identified in this chapter: first, that infrastructure has effects beyond the region in which it is located; second, that adequate assessment of infrastructure requires understanding of a complete network; and third, that traditional static regional assessments are inadequate where major changes in infrastructure occur, since a fuller analysis of the way in which transport can be used by relevant sectors of the economy is necessary.

These broad and often unpredictable impacts imply the need for a wider appreciation of policy than can be achieved by regional or even national authorities, whether or not private capital is involved. The asymmetry between regions incurring costs and regions incurring benefits requires either that major infrastructure should be provided only at the highest — that is, EC — level, where all these factors can be taken into account, or that a formal procedure should be instituted to assess inter-regional factors and allow appropriate compensation to be paid.

Examples of the need for this approach can be seen in the failure to cope with the problem London poses for northern British regions, in the need for improved infrastructure to benefit Irish traffic across the UK, and in the need to compensate regions like Hainaut in Belgium, or Picardie in France, which will suffer construction of a high-speed rail line without gaining adequate access to it. As discussed in Chapter 9, it will be important for the EC to play a wider role in defining and securing finance for infrastructure improvements of Community importance.

Part of the problem, as has already been noted for transport infrastructure, is that a particular piece of infrastructure may well serve many levels of user unless, for example, new motorways are built with minimal access to reduce the amount of local traffic, or new railway services have minimal stopping points. However, it may not be possible to discriminate between different classes of traffic, since one of the compensations for the environmental disturbance which a region may face from new infrastructure is that it, too, will have access to it.

In such circumstances, it becomes extremely difficult to formulate a clear and coherent policy for infrastructure, especially in highly congested, complex and possibly border regions. What is, however, clear is that this type of problem is not just a regional or even a national problem, but essentially a Community problem. What is also clear is that a developed and robust methodology for measuring wider transport infrastructure impacts is needed as the basis for a coherent EC transport strategy.

Note

1. Congestion in London alone is estimated to cost Marks and Spencer £2 million a year, J. Sainsbury £3.4 million, British Telecom £7.25 million, and Royal Mail Letters £10.4 million (Confederation of British Industry, 1989b).

11 *The Channel Tunnel and European integration*

In commissioning the Coopers & Lybrand/SETEC Economie study into the feasibility of a Channel fixed link at the end of the 1970s, the European Commission became a prime, if passive, mover in relaunching the project which in 1986 was to become the Eurotunnel scheme. Indeed, interest from the European Parliament and Commission was strong throughout the late 1970s and early 1980s, and although it probably had little impact on the British and French governments' decision to proceed, it may at least have helped ensure that during this period the Channel fixed link remained on policy-makers' agendas.

Euro-enthusiasm for the Channel fixed link is readily explained by the impetus it was thought such a project would give to European integration. The Channel crossing was identified as one of a number of transport bottlenecks in Europe, and resolution of the difficulties it caused was pursued in connection with a series of similar projects, including in particular the Straits of Messina crossing. However, it was clearly more substantial and more visible than all other bottleneck projects, including the Straits of Messina, and had the potential not simply to improve transport links within the EC, but also to provide a potent symbol of the coming together of the Community. In passing, it may be noted that the British government probably shared this ambition with the EC: it, too, saw the Channel fixed link as a means of demonstrating a commitment to the European ideal. The difference is that whereas the EC viewed the Channel fixed link as a means of promoting European integration, the British government saw it as a means of obscuring a lack of enthusiasm for large parts of this process.

Subsequently, the EC has pressed ahead with its policy of integration through improved transport links, not only by offering to finance (in part) key infrastructure investment but also by encouraging bodies such as the Community of European Railways to develop pan-European transport projects.[1]

The purpose of this chapter is to consider the impact of changing transport systems, whether or not conceived at a European level, on the process of European integration. The chapter's main concern is obviously the role of the Channel Tunnel in this process, but other projects are also investigated when necessary. The chapter looks successively at theories of European integration and expectations of the Tunnel's impact in the light of these; at how these theories measure up to what can so far be observed as having happened in the case of the Tunnel; and at a case study of institutional development at the local level, in the Euroregion formed by Kent County Council (KCC) and the Conseil Régional Nord-Pas de Calais (CRNPC), which could be a harbinger of wider institutional change in the EC.

11.1 Theories and expectations of integration

The Channel Tunnel has consistently been hailed as a development of European significance. Both supporters and critics have maintained that it is inextricably linked to economic, political, social and psychological integration in Europe, implicating Britain in particular in these processes. Thus, President Mitterrand (1987) has gone so far as to say that

> the Channel Tunnel . . . is nothing less than a revolution in habits and practices. It is a fact that, if all the European means of communication can be linked up with the Channel Tunnel, then the whole of Community Europe will have one nervous system and no one country will be able indefinitely to run its economy, its society, its infrastructural development independently from the others.

It is clear from this and from other similar statements that the Tunnel is thus seen by many as central to integration in Europe.

Such a perspective is interesting not simply for what it reveals about attitudes to the Channel Tunnel but also for the unspoken assumptions about mechanisms of integration on which it clearly depends. Our first task is, then, to explore in some detail theories of integration, in order to expose these mechanisms to analysis.

Theories of integration

In the years after the last war, debate about integration in Europe was opened by federalists who sought to bring Western Europe together by formal political means. In utopian manner, they imagined that, in a continent recently taken to the verge of destruction by division, public opinion might spontaneously respond to proposals for creation of a European state and central government (Harrison, 1974). When this did not happen, federalist attention turned to more evolutionary processes, whereby the institutions of a European Community would serve as an emergent federal authority in Europe. This perspective has remained influential, though its emphasis on central authority has tended to obscure other aspects of federalism, in particular its concern to preserve diversity in a new form of European unity.

It was, however, challenged at the end of the 1950s by Deutsch, who argued that integration would come not from deliberate political initiative but from a growing commonality of perceptions and values promoted by increased communication and transaction between nations.

The product would be a 'security community', in which peaceful resolution of disputes would be possible, as was increasingly the case in the North Atlantic area (Deutsch *et al.*, 1957). However, just as federalism began to shift its ground with the development of the new institutional structures of the EEC, so transactionalism was even more emphatically eclipsed.

The creation of the EEC, following the success of the ECSC, suggested that European integration was moving forward on a neo-functionalist basis. The argument here was that there is an ineluctable spill-over from integration in one sector of the economy to integration in another. The work of political elites, interest groups and others was seen as likely to lead to the extension of Commission and Community competence, and to the progressive dismantling of national frontiers. Furthermore, the course of integration in the 1950s and early 1960s seemed to some extent to justify these assumptions.

However, the cumulative impact of de Gaulle, of British accession to the Community, and of world economic crisis after 1973, served to call the whole process of integration into question. National sovereignty did not wither away, and low politics in the Community remained compartmentalised and competitive. Indeed, political elites seemed resolutely to resist the political integration which had been thought to be the inevitable product of functional integration, as national bureaucracies defended the national interest and resisted extension of Community activity, at least until the passage of the Single European Act (George, 1985). The consequent stagnation of the Community created a kind of theoretical void, into which ideas of intergovernmentalism and interdependence moved. The conflictual nature of the Community after 1973 suggested to many that the nation state, far from fading away, had grown in strength, and was increasingly acting as gatekeeper in all dealings with Brussels. Gladiatorial confrontation had replaced spill-over in what was seen as a zero-sum game, modelled closely on normal foreign policy processes.

Nevertheless, beneath increasingly visible high-political conflict, interdependent policy-making and interpenetrating administrative elites were gradually emerging. Neo-functionalism had assumed the isolation of policy-making from the European level, unless spill-over took place. In practice, virtually all areas of policy are now vertically integrated with the Community level. The *engrenage* engendered by the Europeanisation of national economies, the role of Community law, the extension of powers by the Single European Act, and generous interpretation of the Treaty of Rome, have all pushed in this direction.

The implications of these developments for theories of integration are uncertain. The EC has not faded into the background of world economic interaction, and it has not been totally subordinated to the wishes of national governments. Yet political integration, and the strong central authority at the centre of a European federation which Mrs Thatcher warned against in her Bruges speech, remain a long way off. In these circumstances, it seems that all theories have something to offer.

Expectations of integration

By supporters and critics alike, the Channel Tunnel was confidently expected to advance the cause of European integration. Indeed, in the early days one major European dimension to the Tunnel was expected to be a direct input from Brussels. That this has not happened is explained in part by the complexity of factors which has also served to confuse the theoretical debate about integration. Not only has there not been a major Tunnel initiative from Brussels, but also the limitations on such a possibility are all too clearly revealed by the Tunnel experience in, for example, the complex bargaining in which the Commission has had to engage to achieve abolition of frontier controls (from which the British government sought to exclude the Tunnel).

Even if the hopes of federalists have been dashed, there is still the chance that increased transactional exchanges promoted by the opening of the Channel Tunnel will increase the pace of integration in Europe. This is possible both at the level of Britain's trade with her EC partners and at the level of personal contact between Britons and other EC nationals. In each case, integration of Britain's transport system into the wider European network can be expected to have an important impact.

Furthermore, other Channel Tunnel impacts must be considered. A study currently being undertaken for DG XVI, the Regional Policy Directorate-General of the European Commission, is concerned to investigate the extent to which the Channel Tunnel may lead not only to integration but also to centralisation (within an increasingly integrated Community). This was the concern of the Alvarez de Eulate Report (European Parliament, 1988b), which noted the historically centralising effects of transport investment, and which sought to suggest not that such investment should therefore not be undertaken, but that accompanying measures to ensure that peripheral regions were not disadvantaged should also be embarked upon. On the basis of the present DG XVI study, it is thus hoped that policies will be adopted both to enhance the benefit of those who are in any case placed to gain from the Tunnel, and to mitigate the costs to those who are likely to lose from it.

What is important, however, in assessing the Tunnel's impact is institutional change. Over the years, British trade and contact with many independent nations has waxed and waned, but it has not always had major institutional implications. The activities of trade delegations and of embassies may have increased, but the structure of British and foreign institutions has not always been affected in significant ways. In the case of integration in Europe, what is sought is not merely increased activity, but structural change in political institutions.

11.2 Experience of the Channel Tunnel

Although some expected the European Commission to play an important role in fixed-link policy-making, through direct subsidy and associated infrastructural grants, its role to date has been relatively limited. As has already been noted, the European Investment Bank (EIB) loaned £1 billion, out of an initial (post-Equity III) banking stake of £5 billion, and has since increased its involvement in the project. In addition, grants have been awarded for limited infrastructural improvements on both sides of the Channel. However, these measures do not amount to an important EC presence in the project. This remains true even if the EIB loan is seen, as it is by some, as a hidden government guarantee to the project, for the guarantee could not be made explicit without the express consent of the British and French governments. This reveals the important procedural obstacle placed in the way of Brussels by the two governments' tight control of fixed-link policy and, moreover, by their decision to concede the link to private enterprise.

In these circumstances, grants from Brussels have only been awarded through standard mechanisms, and then on a limited scale. They have been made available for transport projects, such as, in the UK, the M20 'missing link' between Maidstone and Ashford, and, in France, the A26 and E20. It is also said that the Channel Tunnel rail link could tap EC funds, if it were not for the British government's refusal to countenance public subsidy of any kind for the link. On a broader scale, although Kent is not eligible on its own for finance through EC Structure Funds, it has been awarded aid through other channels. Thus, in November 1987, the districts of Dover and Thanet were awarded priority status under European Coal and Steel Community provisions, thereby making them eligible for low-cost loans for measures to offset the run-down in the Kent coalfield. Additionally, an ECSC loan of £500 000 has already been made available to help develop RAF Manston as Kent International (civil) airport. In May 1988, the same part of East Kent became eligible for priority funding from the European Social Fund. More importantly, in Autumn 1990, Kent and Nord-Pas de Calais were joint recipients of £14 million awarded to the Transmanche 'Euroregion'.[2]

Beyond this, EC policy has been to facilitate Channel Tunnel progress as much as possible. The Commission has agreed that for at least three years it will approve the utilisation agreement of 29 July 1987, despite its apparent conflict with EC competition policy. It has also accepted that there should not be any interference with stringent British controls required to prevent the spread of rabies. In addition, as mentioned above, it has started to commission research into the impacts of the Tunnel through DG XVI.

This activity is clearly in line with the wishes of the European Parliament, which is keen to develop a more active Channel Tunnel policy. In the summer of 1988, the Parliament considered two reports supporting the project, but calling for aid to offset job losses in the ferry industry and elsewhere, for the abolition of border controls, and for the provision of supporting infrastructure. The Marshall Report (European Parliament, 1988a) recognised the construction of a fixed link between Britain and the continental EC as an event of major political importance, and noted the expectation that the Tunnel would improve human and geographical contacts and mobility. It held that the Tunnel would make the concept of European integration more tangible to citizens of the Community. This Report was approved by Parliament in July 1988, and may be taken as a vote of confidence in the competence and flexibility of European firms to handle a project of this magnitude, and in the new high-speed rail network which was to be developed in Europe.

The Alvarez de Eulate Report stressed the need for social measures to be taken in conjunction with the Tunnel. Looking at the Channel crossing in conjunction with that of the Straits of Messina, it called for detailed research into the economic and social impact of the Tunnel, and for the development both of regional policy to deal with identified problems, and for EIB assistance where necessary. A resolution was adopted on this basis by the full Parliament, despite continuing reservations on the part of some British Labour MEPs, who remained concerned about the possible negative impact on jobs. As was noted above, DG XVI has now commissioned research into the regional impact of the Channel Tunnel.

Thus, although there has hardly been an aggressive federalist response to the Channel Tunnel on the part of the EC, largely because there could not be, there has been a degree of Community interest and involvement which suggests that a mere increase in transactional exchanges is not the only 'European' impact to be expected of a Channel fixed link. Furthermore, important supranational progress is being made outside the confines of the EC itself, by such bodies as the Community of European Railways, and the European Conference of Ministers of Transport. It may be, however, that yet more important and significant moves towards European integration are being made not at the supra- but at the subnational level, where genuine institutional innovation and change may be observed.

11.3 The Kent/Nord-Pas de Calais 'Euroregion'

In view of the role national frontiers play in depressing the economic performance of adjacent regions,[3] one of the more interesting EC initiatives is its trans-frontier region programme. This attempts to enable co-ordinated development to take place in adjacent regions in different member states, and may lead eventually to constitution of 'Euroregions', comprising parts of more than one nation state.

As has already been noted, KCC and the CRNPC have come forward with a number of independent proposals for Community aid to help offset job losses, particularly in the port and ferry industries, that are likely to result from Channel Tunnel operation, and to develop associated transport infrastructure. In addition, in the late 1980s they submitted a joint bid to Brussels, under Article 10 of the recent Regional Development Fund regulations, which governs transfrontier development projects. The Community response was initially favourable, though rather sceptical. A series of discussions between local authority and

Community officials led the Commission, on 6 May 1988, to invite KCC and CRNPC to submit a formal transfrontier project to the EC. In January 1989, the process was further developed when the Commission formally established a Joint Development Concept in the area delimited by the two authorities. At this point, the Commission invited KCC and CRNPC to determine what a genuine Euroregion might look like. Funding for necessary research was provided jointly by the Commission and the two authorities, and work progressed in 1989 and 1990. The Transmanche Euroregion has now been officially recognised by Brussels, and, as was noted above, has already secured EC finance. The process which resulted in the establishment of the Transmanche region was clearly prompted by the decision to build a Channel Tunnel between Kent and Nord-Pas de Calais, and requires further investigation.

In essence, the transfrontier development programme derives from a tourism initiative linking a regional park around Boulogne (the parc naturel régional du Boulonnais) with four Kent districts (Canterbury, Dover, Shepway and Thanet), which was signed at Le Wast (Pas de Calais) on 11 June 1986. Previously to this, contacts existed between KCC and CRNPC, but they were not extensive, and they had not resulted in co-operation agreements. In the Le Wast agreement, specific mention was made of the Channel Tunnel, and of the two authorities' intention to address jointly the problems it might raise. Official inauguration of political links, based on shared experience of the Channel Tunnel, soon followed, on 17 June 1986. At this point, regular meetings between Tony Hart, leader of KCC, and Noël Josèphe, president of CRNPC, were instituted, on a bilateral basis, and with the British and French co-chairmen of Eurotunnel. Furthermore, a joint approach to Brussels for EC funds was agreed at this time. Subsequently, a training initiative was added to the tourism agreement, which was itself enlarged to encompass cultural exchanges and joint marketing of the two subregions.

Genuine progress was made when Hart and Josèphe signed a *protocole d'accord* at Leeds Castle, near Maidstone, on 24 April 1987. This formed the basis for development of a Kent/Nord-Pas de Calais 'Euroregion', in that it pledged the two authorities to work not just in co-operation, but also in common, through joint meetings of officials. It also sought to delimit more clearly a joint programme area which would form the basis for submissions to Brussels. This came to be defined as the five East Kent districts of Ashford, Canterbury, Dover, Shepway and Thanet, and the four French 'districts' of Boulogne-Montreuil, Calais, Dunkerque and St Omer. Finally, it made explicit the strategy of a joint approach to the EC.

At the heart of this agreement was institution of a joint commission, under the co-presidency of Hart and Josèphe, charged with directing the affairs of the Euroregion, and having its own secretariat and the ability to establish working groups when necessary. A number of major areas of common interest were identified, including infrastructure, training, and the joint promotion of tourism. Tourism is an interesting case, because joint promotion can be extremely cost-effective. However, perhaps more interesting in terms of the debate which has been conducted here are the other two major areas of joint action — infrastructure (especially transport) and labour through training. As has already been argued, the co-ordination of infrastructure investment on the two sides of a national frontier is clearly vital, whether or not the separation is physical (as in the case of the Channel). In the circumstances of the single market after 1992, when cross-border movement of factors of production is likely to be greatly increased, it is yet more important to avoid major daily movements which mean that the benefits of new developments in one region largely accrue to the residents of a neighbouring region. This is, in part, why the joint training initiative is important.

The Transmanche Euroregion is still a relatively recent initiative, and in many senses an interesting institutional innovation, the broader ramifications of which are yet to be determined. To some, it is merely a cynical attempt by two EC regions (one of which does not have EC Structure Funds designation) to tap Community resources. Undoubtedly, there is much to be said for this view. To others, it is a measure of grassroots integration which points the way for the Community as a whole.

To proponents of this view, it has to be said that the Euroregion remains a distinctly administrative phenomenon. 'Integration' has not really gone far beyond the periodic meetings of a few high-ranking local government officials, though it should be noted that these officials have themselves moved a long way in opening contacts with officials in Brussels, and have thereby developed a degree of local authority 'autonomy' from British and French central government. Outside these official Anglo-French contacts, practical change is so far limited. The local newspapers of the two regions, the *Kent Messenger* and *La Voix du Nord*, have started to collaborate, and will periodically print copies of *Rendez-vous!* for distribution with their respective papers, but there is little sense yet of identification with a Euroregion. Clearly, other transactional exchanges have also increased, notably in the property market as a result of strong British interest in cheap French houses. The overall impression is, however, of little Anglo-French engagement, and of even less awareness of the new institutional links and structures which are being forged by Kent and Nord-Pas de Calais.

However, it is in many ways premature to judge the success of the Euroregion. Furthermore, it is perhaps only with the existence of a physical link between Kent and Nord-Pas de Calais that progress can be expected. Once the Channel Tunnel is operational, there is a good chance that its high visibility will serve as a focal point for the Euroregion, and that genuine integration will proceed. At this point, institutional change in Kent and Nord-Pas de Calais may become a model for other parts of Europe to adopt.

11.4 Conclusion

Specific examples of increased transactional exchange as a result of Channel Tunnel construction may readily been given. Through the DTI's *Channel Fixed Link Newsletter*, British contractors are now made aware of large construction projects throughout Europe, such as the Great Belt link in Scandinavia; and contact between Kent and Nord-Pas de Calais has clearly increased in recent years. This is perhaps the necessary preliminary to institutional change, and should not, therefore, be omitted from any account which adopts a rather sceptical line concerning existing institutional change. However, a wider perspective needs to be adopted if the full significance of the Channel Tunnel is to be captured.

In many respects, the Tunnel is a project of European significance. Now largely an accepted part of the European landscape, it has been investigated in this book chiefly in terms of its economic and political implications. Its broader role in integrating rather reluctant Britons into the European Community has yet to be properly tested, though the emergent Transmanche Euroregion is perhaps an indication of things to come. In this sense, as in many others detailed in this book, its historic importance should not be underestimated.

Notes

1. See Chapter 9.
2. See below pp.194–6.
3. See Chapter 10.

Bibliography

Agence d'Urbanisme et d'Aménagement de la Région Dunkerquoise, 1973, *Rapport de Présentation du Schéma Directeur d'Aménagement et d'Urbanisme de la Région Dunkerquoise* (AGUR, Dunkerque).

Ashford, D., 1982, *British Dogmatism and French Pragmatism: Central–Local Policymaking in the Welfare State* (Allen & Unwin, London).

Ashford, D., 1989, 'British Dogmatism and French Pragmatism Revisited' in C. Crouch and D. Marquand, *The New Centralism: Britain out of Step in Europe* (Blackwell, Oxford) 77–93.

Atkins, W.S., 1989, *The Impact of the Channel Tunnel on Wales* (Institute of Directors, Wales, Cardiff).

Balassa, B. and L. Bauwens, 1988, 'The Determinants of Intra-European Trade in Manufactures', *European Economic Review* 32, 1421–37.

Bendixson, T., 1989, *Transport in the Nineties: The Shaping of Europe* (Royal Institution of Chartered Surveyors, London).

Biehl, D., 1986, *The Contribution of Infrastructure to Regional Development*, Final Report of the Infrastructure Study Group (Commission of the European Communities, Luxembourg).

Biehl, D., 1991, 'The Role of Infrastructure in Regional Development', in R.W. Vickerman (ed.), *Infrastructure and Regional Development* (Pion, London).

Birnbaum, P., 1982, *The Heights of Power: An Essay on the Power Elite in France* (University of Chicago Press, Chicago).

Blum, U., 1982, 'Effects of Transportation Investments on Regional Growth: A Theoretical and Empirical Investigation', *Papers and Proceedings of the Regional Science Association* 49, 169–84.

Blum, U., 1991, 'The New East–West Corridor: An Analysis of Passenger Transport Flows inside of and through Germany in 2010' in R.W. Vickerman (ed.), *Infrastructure and Regional Development* (Pion, London).

Bonavia, M., 1987, *The Channel Tunnel Story* (David & Charles, Newton Abbot).

Bonnafous, A., F. Plassard and D. Soum, 1975, *Impact of Infrastructural Investment on Industrial Development*, Round Table 25 (European Conference of Ministers of Transport, Paris).

Botham, R., 1983, 'The Road Programme and Regional Development: The Problem of the Counter-factual' in K.J. Button and D. Gillingwater (eds), *Transport, Location and Spatial Policy* (Gower, Aldershot).

British Railways Board, 1989, *International Rail Services for the United Kingdom* (British Railways Board, London).

Bröcker, J. and K. Peschel, 1988, 'Trade' in W. Molle and R. Cappellin (eds), *Regional Impact of Community Policies in Europe* (Avebury, Aldershot), 127–51.

Bruce-Gardyne, J. and N. Lawson, 1976, *The Power Game: An Examination of Decision-making in Government* (Macmillan, London).

Brunet, R., 1989, *Les Villes européennes* (DATAR/La Documentation Française, Paris).

Bulpitt, J., 1989, 'Walking Back to Happiness? Conservative Party Governments and Elected Local Authorities in the 1980s' in C. Crouch and D. Marquand, *The New Centralism: Britain out of Step in Europe* (Blackwell, Oxford) 56–73.

Cagnato, A., 1990, 'Town and Country Planning and Transport Planning Issues in EEC Trans-border Areas: The Case of the North-eastern Regions of Italy' in *Proceedings of Seminar E: Land Use Planning in Europe*, PTRC, Summer Annual Meeting, Brighton.

Cappellin, R., 1990, 'The European Internal Market and the Internationalisation of Small- and Medium-sized Enterprises' in R. Vickerman (ed.), *1992 and Regional Development in Europe, Built Environment* 16, 69–84.

CEDRE, 1990, *Transports à grande vitesse: Développement régional et aménagement du territoire*, rapport de synthèse (Centre Européen du Développement Régional, Strasbourg).

Centre for Local Economic Strategies, 1989, *Channel Tunnel: Vicious Circle*, Research Studies Series No 2 (CLES, Manchester).

Chambre Régionale de Commerce et d'Industrie Nord-Pas de Calais, 1989, *Le Nord-Pas de Calais en quelques chiffres* (CRCINPC, Lille).

Channel Tunnel Joint Consultative Committee, 1986, *Kent Impact Study: A Preliminary Assessment* (CTJCC, London).

Channel Tunnel Joint Consultative Committee, 1987a, *Kent Impact Study: Consultation Document* (CTJCC, London).

Channel Tunnel Joint Consultative Committee, 1987b, *Kent Impact Study: Overall Assessment* (HMSO, London).

Channel Tunnel Joint Consultative Committee, 1989, *Kent Impact Monitoring: First Report of the Channel Tunnel Impact Monitoring Group* (Kent County Council, Maidstone).

Channel Tunnel Joint Consultative Committee, 1990, *Kent Impact Study: Second Monitoring Report* (Kent County Council, Maidstone).

Channel Tunnel Safety Authority, 1990, *Non-Segregation of Drivers and Passengers from their Vehicles* (CTSA, London & Paris).

Chaplain, C., 1990, *L'Insertion de la dynamique transmanche dans le développement du Calaisis: utopie ou réalité* (INRETS/TRACES, Villeneuve d'Ascq).

Colin Buchanan and Partners, 1989, *The Case for a Channel Tunnel Terminal at Stratford: A Report to the London Borough of Newham.*

Commission of the European Communities, 1980, *The Nature and Extent of a Possible Community Interest in the Construction of a Fixed Link across the Channel*, Some Results of a Study for the EEC by Coopers Lybrand Associates and SETEC-Economie (European Commission, Brussels).

Commission of the European Communities, 1986a, *Medium Term Transport Infrastructure Programme*, Communication from the Commission to the Council, COM(86)340 (European Commission, Luxembourg).

Commission of the European Communities, 1986b, *Towards a European High-Speed Rail Network* COM (86)341 (European Commission, Luxembourg).

Commission of the European Communities, 1987, *The Regions of the Enlarged Community: Third Periodic Report on the Social and Economic Situation and Development of the Regions of the Community* (European Commission, Luxembourg).

Commission of the European Communities, 1988a, *Research on the Costs of Non-Europe* (European Commission, Luxembourg).

Commission of the European Communities, 1988b, *Proposal for a Council Regulation for an Action Programme in the Field of Transport Infrastructure*

with a view to the Completion of an Integrated Transport Market in 1992, COM(88)340 (European Commission, Luxembourg).
Commission of the European Communities, 1989, *Transport Infrastructure Policy — Concentration of Efforts and Means*, Communication from the Commission to the Council, COM(89)238 (European Commission, Luxembourg).
Community of European Railways, 1989, *Proposals for a European High-Speed Network*.
Confederation of British Industry, 1989a, *Transport in London: The Capital at Risk* (CBI, London).
Confederation of British Industry, 1989b, *Trade Routes to the Future: Meeting the Transport Infrastructure Needs of the 1990s* (CBI, London).
Conference of Regions of North West Europe (CRONWE), 1989, *Strategic Transport Axis in NW Europe*, ILS Schriften 31 (Institut für Landes- und Stadtentwicklungsforschung des Landes Nordrhein-Westfalen, Dortmund).
Conseil Régional Nord-Pas de Calais, 1985, *Impacts et perspectives pour la région Nord-Pas-de-Calais du lien fixe transmanche*, rapport présenté par Bechtel France (CRNPC, Lille).
Conseil Régional Nord-Pas de Calais, 1986a, *Lien fixe transmanche: éléments pour un plan de développement de la région Nord-Pas de Calais*, dossier remis à Monsieur François Mitterrand, Président de la République, par Monsieur Noël Josèphe, Président du Conseil Régional (CRNPC, Lille).
Conseil Régional Nord-Pas de Calais, 1986b, *La Liaison-fixe transmanche: hier, aujourd'hui, demain* (CRNPC, Lille).
Conseil Régional Nord-Pas de Calais, 1987, *Le TGV Nord* (CRNPC, Lille).
Cornwall County Council, Devon County Council and Plymouth City Council, 1989, *The Channel Tunnel Impact Study* (County Engineers and Planning Department, Devon County Council, Exeter).
Corporation of London, 1990, *London's Transport: A Plan to Protect the Future* (Segal Quince Wicksteed/MVA, London).
Craven, J., 1988, 'The Channel Tunnel: Competition with the Ferries' in Channel Tunnel Research Unit, *Policy Responses to the Channel Tunnel* (CTRU, University of Kent, Canterbury) 151–9.
Crozier, M., 1964, *The Bureaucratic Phenomenon* (University of Chicago Press, Chicago).
Cutler, T., C. Haslam, J. Williams and K. Williams, 1989, *1992: The Struggle for Europe* (Berg, Oxford).
Department of the Environment, 1973a, *The Channel Tunnel Project*, Cmnd. 5256 (HMSO, London).
Department of the Environment, 1973b, *The Channel Tunnel: A UK Cost-Benefit Study*, report by Coopers and Lybrand Associates (HMSO, London).
Department of the Environment, 1973c, *The Channel Tunnel*, Cmnd. 5430 (HMSO, London).
Department of the Environment, 1975, *The Channel Tunnel and Alternative Cross-Channel Services*, a report to the Secretary of State for the Environment by the Channel Tunnel Advisory Group (Chairman, Sir Alec Cairncross) (HMSO, London).
Department of Transport, 1977, *Report of the Advisory Committee on Trunk Road Assessment*, Chairman, Sir George Leitch, (HMSO, London).
Department of Transport, 1982, *Fixed Channel Link: Report of UK/French Study Group*, Cmnd. 8561 (HMSO, London).
Department of Transport, 1986a, *The Channel Fixed Link*, Cmnd. 9735 (HMSO, London).

Department of Transport, 1986b, *The Channel Fixed Link: Concession Agreement*, Cmnd. 9769 (HMSO, London).
Department of Transport, 1990, *Trunk Roads, England: Into the 1990s* (HMSO, London).
Deutsch, K. *et al.*, 1957, *Political Community and the North Atlantic Area: International Organisation in the Light of Historical Experience* (Princeton University Press, Princeton, NJ).
Direction Départementale de l'Équipement du Pas-de-Calais, 1988, *Le Plan routier transmanche* (Conseil Général du Pas-de-Calais, Arras).
Dormard, S., 1988, 'Essai de modelisation du marché de transport transmanche' in Channel Tunnel Research Unit, *Policy Responses to the Channel Tunnel* (CTRU, University of Kent, Canterbury) 140–50.
Duchêne, G., 1988, 'Faudra-t-il réglementer le Tunnel sous la Manche?' in *Lien Fixe Transmanche*, Cahiers du CRAPS, numéro hors-série (CRAPS, Université de Lille II), 156–66.
Dufau, J., 1979, *Les Concessions de service public* (Moniteur, Paris).
Dufloux, C. and L. Margulici, 1984, *Les Eurocrédits: Pourquoi? Comment?* (Editions Banque, Paris).
Dupuit, J., 1844, 'On the Measurement of the Utility of Public Works', *Annales des Ponts et Chaussées* 8; reprinted in D. Munby, *Transport* (Penguin, London, 1968).
The Economist, 1989, 'Allez-oop, Angleterre', 19 August, p.24.
The Economist, 1990, 'The Scandinavian Connection', 28 April, p.65.
Economist Intelligence Unit, 1989, *European Trends* 3.
Emerson, M., 1988, *The Economics of 1992* (Oxford University Press, Oxford).
The European, 1990, 'Stick and Carrot to Build the Tunnel', 11–13 May, p.19.
European Parliament, 1981, *Report on behalf of the Committee on Transport on the Construction of a Channel Tunnel* (rapporteur P. De Keersmaeker), Document 1-93/81.
European Parliament, 1988a, *Report Drawn up on Behalf of the Committee on Transport on the Channel Tunnel* (rapporteur J Marshall), Document A2-328/87.
European Parliament, 1988b, *Report on the Regional Impact of the Construction of a Tunnel under the English Channel and a Bridge over the Strait of Messina* (rapporteur A. de Eulate), Document A2-0102/88.
Eurotunnel, 1987, *Prospectus*.
Eurotunnel, 1990a, *Proposed Financing Programme: Project Information and Notice of Shareholder Meetings*.
Eurotunnel, 1990b, *Eurotunnel Rights Issue*.
Fayard, A., 1980, *Les Autoroutes et leur financement* (La Documentation Française, Paris).
Ferris, T.P., 1990, 'Some Implications of the Channel Tunnel for Ireland', paper to *Third International Conference on the Impact of the Channel Tunnel on Freight Movements in Europe*.
Financial Times, 1990a, 'Bank Plays Key Role in Tunnel Rescue', 20 February, pp.1 and 22.
Financial Times, 1990b, 'Tunnel Project Chief is Replaced', 22 February, p.10.
Franco-British Channel Link Financing Group, 1984, *Finance for a Fixed Channel Link*, 2 vols. (London).
Gamble, A., 1988, *The Free Economy and the Strong State: The Politics of Thatcherism* (Macmillan, Basingstoke).
George, S., 1985, *Politics and Policy in the European Community* (Clarendon Press, Oxford).

Gerardin, B., 1990, 'France', in European Conference of Ministers of Transport, *Private and Public Investment in Transport*, Round Table 81 (ECMT, Paris) 5–32.

Guillot, M., 1988, *Guide expérimental du développeur: grands chantiers et conversion* (Syros, Paris).

Gunnell, J., 1988, 'A Minimum Investment Programme for the North', *Town and Country Planning*, May, 153–5.

Gwilliam, K., 1979, 'Transport Infrastructure Investment and Regional Development' in J.K. Bowers (ed.), *Inflation, Development and Integration* (Leeds University Press, Leeds) 241–62.

Gwilliam, K., 1983, 'Appraisal of the Channel Tunnel Project' in K.J. Button and A. Pearman (eds), *The Practice of Transport Investment Appraisal* (Gower, Aldershot) 88–107.

Haining, P., 1989, *Eurotunnel: An Illustrated History of the Channel Tunnel Scheme*, first published by New English Library in 1973 (Channel Tunnel Group Limited, Folkestone).

Hall, P., 1980, *Great Planning Disasters* (Weidenfeld & Nicolson, London).

Harrison, A., 1974, *The Economics of Transport Appraisal* (Croom Helm, London).

Hayward, J., 1974, 'National Aptitudes for Planning in Britain, France and Italy', *Government and Opposition* 9, 397–410.

Hayward, J., 1976, 'Institutional Inertia and Political Impetus in France and Britain', *European Journal of Political Research* 4, 341–59.

Hayward, J., 1982, 'Mobilising Private Interests in the Service of Public Ambitions: The Salient Element in the French Policy Style?' in J. Richardson (ed.), *Policy Styles in Western Europe* (Allen & Unwin, London) 111–40.

Hayward, J., 1986, *The State and the Market Economy: Industrial Patriotism and Economic Intervention in France* (Wheatsheaf, Brighton).

Heclo, H. and A. Wildavsky, 1981, *The Private Government of Public Money: Community and Policy inside British Politics*, 2nd edn (Macmillan, London).

Henderson, N., 1987, *Channels and Tunnels: Reflections on Britain and Abroad* (Weidenfeld & Nicolson, London).

Henley, A., A. Carruth, A. Thomas and R. Vickerman, 1989, 'Location Choice and Labour Market Perceptions: A Discrete Choice Study', *Regional Studies* 23, 431–45.

HMSO, 1986, *Treaty between the United Kingdom of Great Britain and Northern Ireland and the French Republic concerning the Construction and Operation of Private Concessionaires of a Channel Fixed Link*, 12 February, Cmnd. 9745 (HMSO, London).

Hoffmann, S., 1974, *Decline or Renewal? France since the 1930s* (Viking, New York).

Holliday, I. and R. Vickerman, 1990, 'The Channel Tunnel and Regional Development: Policy Responses in Britain and France', *Regional Studies* 24, 455–66.

Holzmann, R. and A. van der Bellen, 1986, 'Comparative analysis of Financing Methods for Highways in European Countries' in G. Terny and R. Prud'homme (eds), *Le Financement des équipements publics de demain* (Economica, Paris) 155–74.

House of Commons, 1981, *The Channel Link*, Second Report of the Transport Committee, session 1980/81, HC 155 (HMSO, London).

House of Commons, 1985, *The Channel Link*, First Report from the Transport Committee, session 1985/86, HC 50 (HMSO, London).

House of Commons, 1986, *Radioactive Waste*, First Report from the Environment Committee, session 1985/86, HC 191 (HMSO, London).

House of Lords, 1989, *Transport Infrastructure*, 21st Report of the Select Committee on the European Communities, session 1988/89, HL 84 (HMSO, London).

Illeris, S. and L. Jakobsen, 1991, 'Effects of the Fixed Link across the Great Belt' in R.W. Vickerman (ed.), *Infrastructure and Regional Development* (Pion, London).

The Independent, 1989a, 'Airlines Urge EC Intervention', 5 June.

The Independent, 1989b, 'An Expert at Crisis Management', 9 October.

The Independent, 1990, 'Theology Disciple a True Believer', 8 January.

INSEE, 1986a, *Les Industries du Nord-Pas de Calais: Eléments statistiques 1962–1985* (Observatoire Economique du Nord-Pas de Calais, Lille).

INSEE, 1986b, *Profils de l'économie Nord-Pas de Calais* 3 (Observatoire Economique du Nord-Pas de Calais, Lille).

Jordan, A. and J. Richardson, 1987, *British Politics and the Policy Process: An Arena Approach* (Unwin Hyman, London).

Kay, J., A. Manning and S. Szymanski, 1989, 'The Economic Benefits of the Channel Tunnel', *Economic Policy* 8, 211–34.

Kay, J., A. Manning and S. Szymanski, 1990, 'Pricing a New Product: Eurotunnel', *Business Strategy Review* 1, 37–56.

Keeble, D., J. Offord and S. Walker, 1988, *Peripheral Regions in a Community of Twelve Member States* (European Commission, Luxembourg).

Keeble, D., P. Owens, and C. Thompson, 1981, *The Influence of Peripheral and Central Locations on the Relative Development of Regions* (Dept. of Geography, University of Cambridge).

Keeble, D., P. Owens and C. Thompson, 1982, 'Regional Accessibility and Economic Potential in the European Community', *Regional Studies* 16, 419–32.

Kent County Council, 1988, *Kent Reports 1988* (KCC, Maidstone).

Kent County Council, 1989, *Kent Reports 1989* (KCC, Maidstone).

Kent County Council/Conseil Régional Nord-Pas de Calais, 1988, *Channel Tunnel Transfrontier Development Programme* (KCC/CRNPC, Maidstone/Lille).

Kent Economic Development Board, 1988, *Kent: Now Everything Points in Our Direction* (KEDB, Maidstone).

King, A., 1975, 'Overload: Problems of Governing in the 1970s', *Political Studies* 23, 284–96.

Klaassen, L., 1991, 'Traffic Developments and Spatial Developments: The Urban Dimension' in *Transport and Spatial Distribution of Activities*, Round Table 85 (European Conference of Ministers of Transport, Paris).

Laboratoire d'Economie des Transports/Interalp, 1986, *Les Effets socio-économiques du TGV en Bourgogne et Rhône-Alpes* (DATAR/INRETS/OEST/SNCF, Paris).

de Lamothe, A., 1965, 'Ministerial Cabinets in France', *Public Administration* 43, 365–81.

Land Use Consultants, 1986, *Channel Fixed Link: Environmental Appraisal of Alternative Proposals*, a report prepared for Department of Transport (HMSO, London).

Lenel, H., 1988, 'Hard-Coal Mining: A Divided and Protected Market' in H. de Jong, *The Structure of European Industry*, 2nd edn (Kluwer, Dordrecht) 61–79.

Lindblom, C., 1959, 'The Science of "Muddling Through"', *Public Administration Review* 19, 79–88.

London Chamber of Commerce, 1989, *The Impact of the Channel Tunnel on London's Business Community* (LCC, London).

Machin, H. and V. Wright, 1985, 'Economic Policy under the Mitterrand Presidency, 1981–1984: An Introduction' in H. Machin and V. Wright (eds), *Economic Policy and Policy-making under the Mitterrand Presidency 1981–1984* (Pinter, London) 1–43.

Mandrin, J., 1967, *L'Enarchie; ou, les mandarins de la société bourgeoise* (La Table Ronde, Paris).

Mény, Y., 1987, 'France: The Construction and Reconstruction of the Centre, 1945–86', *West European Politics* 10 (iv) 52–69.

Metge, P. and J.-Y. Potel, 1987, *Les Régions du nord de la France et le tunnel sous la Manche* (ACT/DATAR, Paris).

Ministry of Transport, 1963, *Proposals for a Fixed Channel Link*, Cmnd. 2137 (HMSO, London).

Mitterand, F., 1987, The Future of Europe, *The World Today* 43(3) 40.

Monopolies Commission, 1974, *Cross Channel Car Ferry Services*, HC 14, 1974 (HMSO, London).

Monopolies and Mergers Commission, 1988, *BR Board: Network South East*, Cm. 204 (HMSO, London).

Monopolies and Mergers Commission, 1989a, *BR Board: Provincial Services*, Cm. 584 (HMSO, London).

Monopolies and Mergers Commission, 1989b, *Cross-Channel Car Ferries*, Cm. 903 (HMSO, London).

Morris, P. and G. Hough, 1986, *Preconditions of Success and Failure in Major Projects*, Technical Paper no. 3 (Major Projects Association, Oxford).

Morton, A., 1989, 'The Impact of the Tunnel', in Financial Times, *Transport Links with the Continent: Collaboration to Meet the Challenges of Future Growth* (Financial Times, London) Chapter 2.

Myrdal, G., 1957, *Economic Theory and Underdeveloped Regions* (Duckworth, London).

Neuberger, H., 1989, *The Economics of 1992* (Socialist Group, European Parliament).

Neven, D., 1990, 'EEC Integration Towards 1992: Some Distributional Aspects', *Economic Policy* 10, 13–62.

Owen, A.D. and G.D.A. Phillips, 1987, 'The Characteristics of Railway Passenger Demand', *Journal of Transport Economics and Policy* 21, 231–53.

Pieda, 1989, *The Impact of the Channel Tunnel on the Scottish Economy* (Scottish Development Agency, Glasgow).

Plogmann, F., 1980, *Die Bedeutung der Verkehrsinfrastruktur für das regionale Entwicklungspotential*, Beiträge zur Siedlungs- und Wohnungswesen und zur Raumplanung 64, Münster.

Pratten, C., 1988, 'A Survey of the Economies of Scale' in Commission of the European Communities, *Research on the Costs of Non-Europe* (European Commission, Luxembourg) Vol. 2, Chapter 2.

Quinet, E. and G. Morancay, 1989, 'The Toll System on French Motorways' in European Conference of Ministers of Transport, *Systems of Road Infrastructure Cost Coverage*, Round Table 80 (ECMT, Paris).

République Française/Région Nord-Pas de Calais, 1986, *Plan Transmanche: Protocole d'Accord État/Région Nord-Pas de Calais* (Lille).

Rhodes, R., 1986, *The National World of Local Government* (Allen & Unwin, London).

Rhodes, R., 1988, *Beyond Westminster and Whitehall: The Sub-Central Governments of Britain* (Unwin Hyman, London).

Richardson, J., G. Gustafsson and G. Jordan, 1982, 'The Concept of Policy Style' in J. Richardson (ed.), *Policy Styles in Western Europe* (Allen & Unwin, London) 1–16.

Richardson, J.J. and A.G. Jordan, 1979, *Governing under Pressure: The Policy Process in a Post Parliamentary Democracy* (Martin Robertson, Oxford).

Rickard, J., 1990, 'United Kingdom' in *Private and Public Investment in Transport*, Round Table 81 (European Conference of Ministers of Transport, Paris).

Roberts, H., 1990, 'The French Planning Context: Prospects and Procedures for the Developer in the Nord Pas-de-Calais', paper presented to Royal Institution of Chartered Surveyors Conference, April.

Round Table of European Industrialists, 1988, *Need for Renewing Transport Infrastructure in Europe* (Brussels).

Royal Town Planning Institute, 1990, *Impact of the Channel Tunnel on the Regions* (RTPI, London).

Secchi, C., M. Antonioli, A. Pio and L. Resmini, 1991, 'Economic and Social Effects of a Fixed Link on the Strait of Messina', in R.W. Vickerman (ed.), *Infrastructure and Regional Development* (Pion, London).

Seldon, A., 1990, 'The Cabinet Office and Coordination 1979–87', *Public Administration* 68, 103–21.

SERPLAN, 1989, *The Channel Tunnel: Implications for the South East Region*, Report RPC 1470.

Shepherd, G., 1984, 'Industrial Change in European Countries: The Experience of Six Sectors' in A. Jacquemin (ed.), *European Industry: Public Policy and Corporate Strategy* (Clarendon Press, Oxford) 191–214.

Shonfield, A., 1965, *Modern Capitalism: The Changing Balance of Public and Private Power* (Oxford University Press, London).

Spectator, 1987a, 'Sinking Money in the Tunnel', 21 February, p.15.

Spectator, 1987b, 'How Eurotunnel Dug Itself out of the Sand', 8 August, p.21.

Steer Davies & Gleave, 1989, *The Right Tracks to Europe: The Regional and Environmental Impact of the Channel Tunnel* (Transport 2000, London).

Stevens, A., 1980, 'The Higher Civil Service and Economic Policy-making' in P.G. Cerny and M.A. Schain, *French Politics and Public Policy* (Pinter, London) 79–100.

Straszheim, M., 1972, 'Researching the Role of Transportation in Regional Development', *Land Economics* 48, 212–19.

Suleiman, E., 1974, *Politics, Power, and the Bureaucracy in France: The Administrative Elite* (Princeton University Press, Princeton, NJ).

Suleiman, E., 1978, *Elites in French Society: The Politics of Survival* (Princeton University Press, Princeton, NJ).

Thiébault, J.-L., 1986, 'La Politique de recession charbonnière et Charbonnages de France' in *Décentralisation, Énergie, Intervention économique: Niveaux d'activité, Niveaux de pouvoir*, Cahiers du CRAPS 1 (CRAPS, Université de Lille II) 23–59.

Thoenig, 1973, *L'Ere des technocrates: le cas des Ponts et Chaussées* (Les Editions d'Organisation, Paris).

The Times, 1987, 'A Tunneller's Second Front', 27 August, p.8.

Touret, B., 1972, 'La Régime juridique des concessions d'autoroutes', *Actualités juridiques: droit administratif* 372–91.

Vickerman, R.W., 1987a, 'The Channel Tunnel and Regional Development: A Critique of Infrastructure-Led Growth', *Project Appraisal* 2, 31–40.

Vickerman, R.W., 1987b, 'The Channel Tunnel: Consequences for Regional Growth and Development', *Regional Studies* 21, 187–97.

Vickerman, R.W., 1988, 'Economic Policy Responses to the Channel Tunnel' in Channel Tunnel Research Unit, *Policy Responses to the Channel Tunnel* (CTRU, University of Kent, Canterbury) 122–39.

Vickerman, R.W., 1989a, 'After 1992: The South East as a Frontier Region' in M. Breheny and P. Congdon (eds), *Growth and Change in a Core Region* (Pion, London) 87–105.

Vickerman, R.W., 1989b, 'Measuring Changes in Regional Competitiveness; The Effects on International Infrastructure Investments', *Annals of Regional Science* 23, 275–86.

Vickerman, R.W., 1990, 'Whither the Core of Europe in the 1990s?' in *Proceedings of Seminar E: Land Use Planning in Europe*, PTRC, Summer Annual Meeting, Brighton.

Vickerman, R.W., 1991a, 'Transport Infrastructure in the European Community: New Developments, Regional Implications and Evaluation' in R.W.Vickerman (ed.), *Infrastructure and Regional Development* (Pion, London).

Vickerman, R.W., 1991b, 'Other Regions' Infrastructure in a Region's Development' in R.W. Vickerman (ed.), *Infrastructure and Regional Development* (Pion, London).

Vickerman, R.W. and A.D.J. Flowerdew, 1990, *The Channel Tunnel: The Economic and Regional Impact*, Special Report no. 2024 (Economist Intelligence Unit, London).

Vickers, J. and V. Wright, 1988, 'The Politics of Industrial Privatisation in Western Europe: An Overview', *West European Politics* 11 (iv), 1–30.

Vickers, J. and G. Yarrow, 1988, *Privatization: An Economic Analysis* (MIT Press, Cambridge, Mass.).

Ville de Lille, 1989, *Le Centre des gares de Lille: Présentation générale.*

Voix du Nord, 1989, 'Trois ministres, les présidents de TML et Eurotunnel: les états généraux du tunnel', 8 July.

Walberg, L., 1989, 'The Scandinavian Link and the Øresund Fixed Link', paper presented to Council of Europe Conference on Regional Transport, May–June.

Watson, A., 1967, 'The Channel Tunnel: Investment Appraisals', *Public Administration* 45, 1–21.

Wilson, H., 1971, *The Labour Government 1964–1970: A Personal Record* (Weidenfeld & Nicolson, London).

Young, H., 1990, *One of Us: A Biography of Margaret Thatcher* (Macmillan, London).

Zoller, R., 1979, 'La Crise des concessions d'autoroutes', *Revue du droit public et de la science politique en France et à l'étranger* 95, 167–212.

Index